MATRIX ALGEBRA for ENGINEERS

Second Edition

James M. Gere
William Weaver, Jr.

School of Engineering
Stanford University

PWS Engineering
Boston, Massachusetts

PWS PUBLISHERS

Prindle, Weber & Schmidt • Duxbury Press • PWS Engineering •
Statler Office Building • 20 Park Plaza • Boston, Massachusetts 02116

PWS Publishers is a division of Wadsworth, Inc.

Printed in the United States of America

10 9 8 7 6 5 4 3 2 1

Library of Congress Cataloging in Publication Data

Gere, James M.
Matrix algebra for engineers.

Bibliography: p.
Includes index.
1. Matrices. 2. Engineering mathematics.
I. Weaver, William, 1929– II. Title.
TA347.D4G45 1982 512.9'434 82-20631
ISBN 0-534-01274-4

Sponsoring Editor: Ray Kingman
Editing and Production: Project Publishing & Design, Inc.
Cover and interior design: Project Publishing & Design, Inc.
Illustrations and typesetting: Project Publishing & Design, Inc.
Production Services Manager: Stacey C. Sawyer

Preface

MATRIX ALGEBRA has assumed a role of great importance to engineers because of the many useful applications of this branch of mathematics in various fields of engineering and applied science. Subjects such as electric circuit analysis, engineering mechanics, structural analysis, and systems analysis require the use of matrix operations for solving complex problems. Therefore, students of engineering need to obtain a working knowledge of matrix algebra as a part of their basic engineering training. This book is intended to fulfill the needs of such students and others who wish to learn the elements of matrix algebra from a practical point of view. A working knowledge of the subject can be obtained from this book, either through self-study or in conjunction with classroom instruction.

The only mathematical prerequisite for understanding this material is algebra, and the subject matter can be mastered easily by college students at all levels. Methodology is emphasized rather than the mathematical structure of the subject, and all of the developments are presented at an elementary level without formal proofs. The authors have taught this material to engineering students for many years and have found that a thorough grasp of the subject may be attained in a relatively short time.

To aid the student in becoming familiar with the matrix operations that are described, numerous illustrative examples are given throughout the text. Many problems to be solved by the student appear at the ends of the chapters, and answers to most problems are given at the back of the book. In addition, a list of references is provided for those who desire further study from a more theoretical point of view.

This second edition is organized in the same manner as the first, but new material on determinants, matrix inverses, inversion by partitioning, square roots of matrices, solutions of simultaneous equations, and repeated eigenvalues has been added to make the book more useful and complete. Also, throughout the text, explanations have been improved and expanded, examples and problems have been revised, and new problems have been added.

We wish to thank everyone who helped us with this new edition: Professor Thomas R. Kane of Stanford for suggestions of a mathematical nature; Bahman Lashkari-Irvani, our student, for proofreading aid; Curt Cowan of Project Publishing & Design, Inc., for his design and production skills; Ray Kingman, engineering editor of Brooks/Cole, for his advice and reviews; and our wives, Janice Gere and Connie Weaver, for patience and moral support.

James M. Gere
William Weaver, Jr.

Contents

1

...Scalars and Vectors...

1.1. Scalars. Before beginning a discussion of matrix algebra, it is useful to review the concepts of scalars and vectors. A *scalar quantity* is one that possesses only magnitude and hence can be identified by a single number; for this reason the numbers themselves sometimes are referred to as scalar numbers. An example of a scalar quantity (or *scalar*) is the speed of a vehicle, which can be specified completely by a single number (such as 80) provided that the units for the number (such as kilometers per hour) have been specified also. When stating the speed of a vehicle, there is nothing implied as to the direction of the motion, nor whether the speed is increasing or decreasing. Other examples of scalars are the distance between two points in space, the temperature at a particular location in a room, and the number of storage locations in the memory of a computer. In each of these illustrations, a single number is sufficient to specify the quantity; for example, the numbers and their units might be 3 meters, 20 degrees Celsius, and 32,000 storage locations, respectively. Of course, a scalar may be positive, negative, or zero.

1.2. Vectors. There are other quantities of a more complicated nature than scalars that cannot be specified by a single number. This fact leads to the concept of a *vector quantity* (or *vector*). Some of the most familiar vector quantities are encountered in elementary mechanics and include forces, moments, velocities, and accelerations. In general, each of these quantities can be represented by a vector requiring three numbers and an associated coordinate system for its complete specification; hence, they are said to be three-dimensional vectors. As an illustration, consider the vector **F** shown in Figure 1-1. (Note that vector quantities are printed in bold-face type to distinguish them from scalars.) The vector **F** is represented by an arrow having an appropriate length and direction in space. It is apparent that a single number is not sufficient to specify this vector, because both its magnitude and direction must be given.

The *magnitude* of a vector is a scalar quantity denoted by the length of the vector; for example, if the vector **F** in Figure 1-1 is a force vector, its magnitude might be 10 newtons, represented to a suitable scale by the length from the tail to the head of the arrow. Thus, the magnitude always has a nonnegative value, that is, it is either positive or zero, and it is usually denoted by absolute value signs; for instance, the magnitude of **F** is written as $|\mathbf{F}|$.

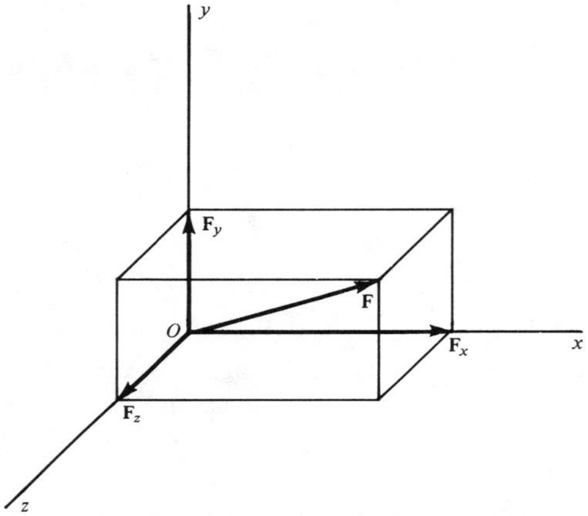

Figure 1-1. A vector **F** and its components

Sometimes it is useful to express a vector in terms of a vector of unit magnitude (called a *unit vector*) that is oriented along the same axis as the vector itself. In order to illustrate this idea, Figure 1-2 shows a unit vector **f** along the axis of the vector **F** of Figure 1-1. When expressed in terms of this unit vector, the vector **F** is written in the form

$$\mathbf{F} = F\mathbf{f} \tag{1-1}$$

in which F denotes the *scalar value* of the vector. If the vector **F** acts in the same direction as the unit vector **f**, the scalar value F will be positive and equal to the magnitude of **F**. However, if the vector is directed along the same axis as **f** but in the opposite direction, then F will be negative. In other words, the sign of the scalar value denotes whether the vector is in the same direction as the unit vector or in the opposite direction. In either case, the magnitude of the vector is equal to the absolute value of the scalar value, or $|\mathbf{F}| = |F|$.

One method of defining the *direction* of a vector in space is to specify the angles between the positive direction of the vector and a set of rectangular coordinate axes x, y, and z. These angles, in combination with the magnitude of the vector, are sufficient to identify the vector without ambiguity. For instance, consider again the vector **F** shown in Figure 1-1. If the three angles between this vector and the x, y, and z axes are given, the positive direction of the vector is established. Then, if the magnitude is known, the vector is completely identified.

However, a more useful way to specify a vector is by means of its components in the x, y, and z directions. The vector components of the vector **F** are shown in Figure 1-1 as $\mathbf{F}_x$, $\mathbf{F}_y$, and $\mathbf{F}_z$. These components can be added vectorially to give the vector **F**; the addition is expressed by the vector equation

$$\mathbf{F} = \mathbf{F}_x + \mathbf{F}_y + \mathbf{F}_z \tag{1-2}$$

The vector components $\mathbf{F}_x$, $\mathbf{F}_y$, and $\mathbf{F}_z$ can be written conveniently in terms of unit vectors **i**, **j**, and **k** oriented in the x, y, and z directions, respectively, as shown in Figure 1-2. Thus, the vectors $\mathbf{F}_x$, $\mathbf{F}_y$, and $\mathbf{F}_z$ can be expressed in the form $F_x\mathbf{i}$, $F_y\mathbf{j}$, and $F_z\mathbf{k}$, respectively, in which F_x, F_y, and F_z represent the scalar values of the component vectors just as F represents the scalar value of the vector itself (see Equation 1-1). Therefore, the vector **F** (see Equation 1-2) can be expressed as

$$\mathbf{F} = F_x\mathbf{i} + F_y\mathbf{j} + F_z\mathbf{k} \tag{1-3}$$

The quantities F_x, F_y, and F_z are called the *scalar components* of the vector **F** in the x, y, and z directions, respectively.

In order to provide a simpler method for expressing the vector, it is convenient to omit the writing of the unit vectors **i**, **j**, and **k** and to state only the scalar components. Thus, the vector **F** can be expressed as

$$\mathbf{F} = F_x, F_y, F_z$$

In this case, it is understood that the terms in the sequence to the right of the equal sign are written in the order x, y, and z, and that each one must be multiplied by the corresponding unit vector in order to obtain the vector components of **F**. To indicate that the sequence of numbers does indeed represent a vector, with the attendant implications, it is customary to enclose them in brackets [ ] and to omit the commas when they are not needed for clarity.

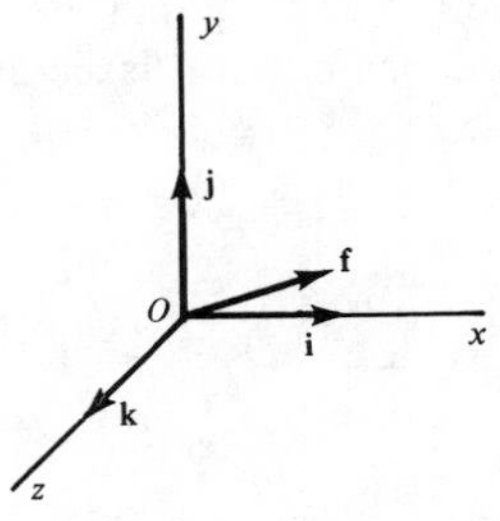

Figure 1-2. Unit vectors

Thus, a useful form in which to represent the vector **F** is

$$\mathbf{F} = [F_x \quad F_y \quad F_z]$$

This expression serves to emphasize that three numbers are required for the specification of a three-dimensional vector, and it also suggests a more general definition of a vector: a vector is an *ordered sequence of numbers*. This definition makes it possible to consider not only the three-dimensional vectors usually encountered in physical problems, but also to consider vectors having any number of dimensions. For example, the sequence of numbers

$$[7 \quad -1 \quad 4 \quad 2 \quad 8]$$

constitutes a five-dimensional vector, and the numbers in the sequence are called the *components* (or *elements*) of the vector. It is seen that the dimension of a vector is the same as the number of its components. There will be a need for vectors of dimension greater than three in subsequent discussions.

In some instances that arise in matrix algebra it is necessary to distinguish between a *row vector*, in which the components are written horizontally in a row, and a *column vector*, in which the components are listed vertically. Thus, in the following illustrations **A** and **B** are row vectors and **C** and **D** are column vectors:

$$\mathbf{A} = [7 \quad 3] \qquad \mathbf{B} = [6 \quad 4 \quad 6 \quad -1 \quad 7]$$

$$\mathbf{C} = \begin{bmatrix} 6 \\ 3 \end{bmatrix} \qquad \mathbf{D} = \begin{bmatrix} 7 \\ 2 \\ -3 \\ 0 \\ 4 \end{bmatrix}$$

There is no intrinsic difference between row and column vectors; the only difference is in the manner of representation. All of the algebraic operations, such as addition, that are described later for row vectors can be performed also with column vectors. Furthermore, a column vector can be printed in a row to save space, with braces { } to indicate that it is a column vector; thus,

$$\mathbf{C} = \begin{bmatrix} 6 \\ 3 \end{bmatrix} = \{6 \quad 3\}$$

$$\mathbf{D} = \begin{bmatrix} 7 \\ 2 \\ -3 \\ 0 \\ 4 \end{bmatrix} = \{7 \quad 2 \quad -3 \quad 0 \quad 4\}$$

This representation of a column vector will often be used for convenience.

1.3. Addition of Vectors. Two or more vectors of the same dimension (the same number of elements) may be added vectorially to give a new vector. For example, Figure 1-3 shows two three-dimensional vectors $\mathbf{F}_1$ and $\mathbf{F}_2$, each of which has a general orientation in space. The sum of $\mathbf{F}_1$ and $\mathbf{F}_2$, denoted as $\mathbf{F}_R$ in the figure, is called the *vector sum* or *resultant vector*. The addition may be accomplished by adding corresponding components of the vectors. Thus, if it is assumed that the vectors $\mathbf{F}_1$ and $\mathbf{F}_2$ are expressed in the form

$$\mathbf{F}_1 = [F_{1x} \quad F_{1y} \quad F_{1z}]$$

$$\mathbf{F}_2 = [F_{2x} \quad F_{2y} \quad F_{2z}]$$

then the sum of the two vectors will be the vector $\mathbf{F}_R$ given as

$$\mathbf{F}_R = \mathbf{F}_1 + \mathbf{F}_2 = [F_{1x} + F_{2x} \quad F_{1y} + F_{2y} \quad F_{1z} + F_{2z}]$$

This equation can be considered as an expression of the parallelogram law for combining vectors in three dimensions. However, the definition can be generalized so that it applies to vectors of any dimension, as shown in the following example for the addition of two five-dimensional vectors:

$$\mathbf{A} = [7 \quad -1 \quad 4 \quad 2 \quad -8] \tag{1-4a}$$

$$\mathbf{B} = [1 \quad 2 \quad 9 \quad 0 \quad -4] \tag{1-4b}$$

$$\mathbf{C} = \mathbf{A} + \mathbf{B} = [8 \quad 1 \quad 13 \quad 2 \quad -12] \tag{1-4c}$$

A similar procedure is followed for the addition of column vectors.

The subtraction of two vectors is carried out in a manner analogous to that for addition; thus, the difference of the vectors **A** and **B** given in Equations (1-4) is

$$\mathbf{D} = \mathbf{A} - \mathbf{B} = [6 \quad -3 \quad -5 \quad 2 \quad -4]$$

It may be seen from these examples that in order to be *conformable* for addition or subtraction, the vectors must have the same dimension, that is, the same number of elements.

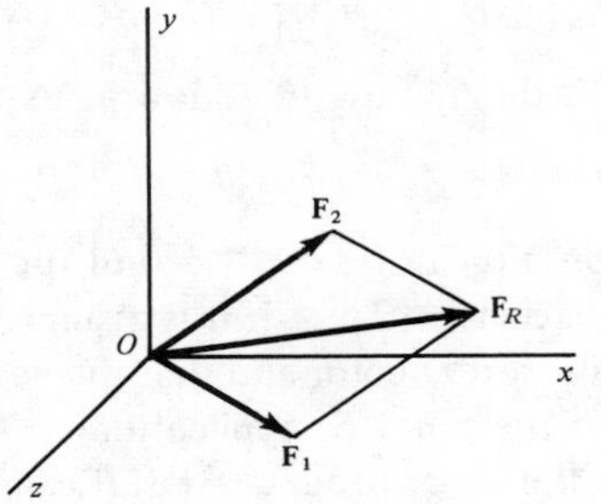

Figure 1-3. Addition of vectors

Occasionally it is necessary to have a *zero vector* for use in computations. This vector, which is also called the *null vector*, contains only zero elements; thus,

$$\mathbf{0} = [0 \quad 0 \quad \ldots \quad 0] \tag{1-5}$$

The null vector performs the same role in vector algebra as does zero (0) in scalar algebra. For example, if the sum of two vectors **A** and **B** is a null vector, the corresponding vector equation is

$$\mathbf{A} + \mathbf{B} = \mathbf{0}$$

in which **A**, **B**, and **0** must have the same dimension.

The addition of a vector to itself produces a new vector in which the scalar value and all elements are twice those of the original vector; thus:

$$\mathbf{A} + \mathbf{A} = 2\mathbf{A} = [2A_1 \quad 2A_2 \quad \ldots \quad 2A_n]$$

in which $A_1, A_2, \ldots, A_n$ represent the components of **A**. In general, successsive additions of the same vector result in a new vector which is the original vector multiplied by a scalar. This process may be accomplished directly by multiplying each element of the original vector by the same scalar. Thus, the notion of multiplying a vector by a scalar derives from the process of successive additions. If λ is any scalar number and **A** is any n-dimensional vector, the product $\lambda\mathbf{A}$ is defined as

$$\lambda\mathbf{A} = [\lambda A_1 \quad \lambda A_2 \quad \ldots \quad \lambda A_n] \tag{1-6}$$

This rule for *multiplication by a scalar* applies to either row or column vectors.

1.4. Multiplication of Vectors. Two kinds of products that can be formed by multiplying vectors are described in the following discussion. The first of these products is the *scalar product* (often called the *dot product*). The scalar product of two vectors **A** and **B** is written in the form $\mathbf{A}\cdot\mathbf{B}$, and it is this representation that has given rise to the name "dot product." The term "scalar product" comes from the fact that the multiplication produces a scalar result.

If two three-dimensional vectors **A** and **B** are given as follows,

$$\mathbf{A} = A_x\mathbf{i} + A_y\mathbf{j} + A_z\mathbf{k} = [A_x \quad A_y \quad A_z] \tag{1-7a}$$

$$\mathbf{B} = B_x\mathbf{i} + B_y\mathbf{j} + B_z\mathbf{k} = [B_x \quad B_y \quad B_z] \tag{1-7b}$$

then their scalar product C is defined by the following expression:

$$C = \mathbf{A}\cdot\mathbf{B} = A_xB_x + A_yB_y + A_zB_z \tag{1-8}$$

As an example, this relationship can be used to find the work of a force that is moved through a given displacement. Thus, if it is assumed that the vector **A** (see Equation 1-7a) is a constant force vector and that the vector **B** (Equation 1-7b) defines the displacement of the point of application of **A**, then the work done by the force **A** is given by the scalar product $\mathbf{A}\cdot\mathbf{B}$. It can be seen from Equation

(1-8) that the scalar product is commutative; that is, $\mathbf{A} \cdot \mathbf{B} = \mathbf{B} \cdot \mathbf{A}$.

The scalar product is used also in determining the scalar component of a vector in a particular direction. This operation is accomplished by taking the scalar product of the given vector and a unit vector having the direction of the desired component. For example, consider the scalar products of the vector **A** defined by Equation (1-7a) with the unit vectors **i**, **j**, and **k** (see Figure 1-2). Since the unit vectors may be written in the form

$$\mathbf{i} = [1 \quad 0 \quad 0] \qquad \mathbf{j} = [0 \quad 1 \quad 0] \qquad \mathbf{k} = [0 \quad 0 \quad 1] \tag{1-9}$$

it follows from Equation (1-8) that the scalar products are

$$\mathbf{A} \cdot \mathbf{i} = A_x \qquad \mathbf{A} \cdot \mathbf{j} = A_y \qquad \mathbf{A} \cdot \mathbf{k} = A_z$$

Thus, the scalar product of **A** with the unit vector **i** gives the scalar component of **A** in the x direction, and similarly for the y and z directions.

It should be noted that if two three-dimensional vectors are perpendicular to one another (that is, *orthogonal*) their scalar product is zero. This conclusion can be seen by assuming that **A** and **B** (Equations 1-7) are orthogonal vectors with the x axis in the direction of **A** and the y axis in the direction of **B**. This choice results in the following values for certain of the components:

$$A_y = A_z = 0 \qquad B_x = B_z = 0$$

Therefore, the scalar product given by Equation (1-8) is zero.

On the other hand, if the two vectors **A** and **B** are in the same direction, their scalar product is equal to the product of their scalar values. One may visualize this result by again considering the vectors **A** and **B** in Equations (1-7) and by assuming that both vectors have the same direction, selected arbitrarily as the x direction. The scalar product of these vectors then consists of the product $A_x B_x$ (see Equation 1-8).

The unit vectors **i**, **j**, and **k** are orthogonal to one another; therefore, the following products are zero:

$$\mathbf{i} \cdot \mathbf{j} = \mathbf{i} \cdot \mathbf{k} = \mathbf{j} \cdot \mathbf{k} = 0 \tag{1-10}$$

Also, since these vectors have unit lengths, the scalar product of any one of them with itself is equal to unity:

$$\mathbf{i} \cdot \mathbf{i} = \mathbf{j} \cdot \mathbf{j} = \mathbf{k} \cdot \mathbf{k} = 1 \tag{1-11}$$

Another way to represent the scalar product of two three-dimensional vectors **A** and **B** (see Figure 1-4) is as follows:

$$\mathbf{A} \cdot \mathbf{B} = |\mathbf{A}|\,|\mathbf{B}| \cos \alpha \tag{1-12}$$

in which $|\mathbf{A}|$ and $|\mathbf{B}|$ are the magnitudes of the vectors **A** and **B** and α is the smaller angle between them. This equation can be interpreted to mean that the scalar product is equal to the magnitude of vector **A** times the magnitude of the

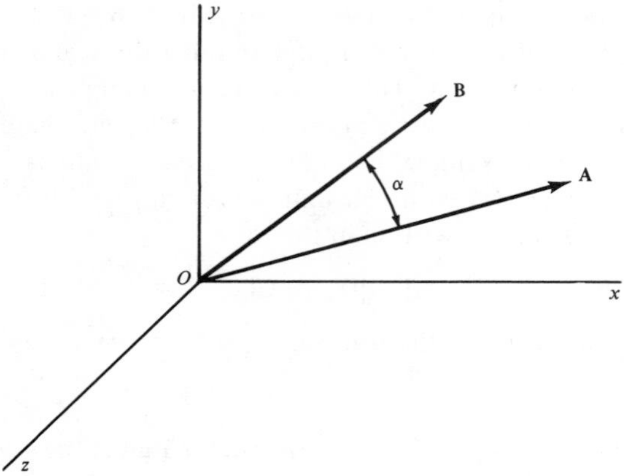

Figure 1-4

projection of **B** in the direction of **A**. Alternatively, this product may be interpreted as the magnitude of **B** times the projection of **A** in the direction of **B**. In the case of orthogonal vectors, the angle α is 90 degrees (or $\pi/2$ radians), and the scalar product obtained from Equation (1-12) is zero. In the case of collinear vectors with their arrows pointing in the same direction, the angle α is zero and the scalar product is equal to the product of their magnitudes. If they are collinear but in opposite directions, the angle is 180 degrees (or π radians), and the scalar product is the negative of the product of their magnitudes. It is left as an exercise at the end of the chapter to show that the two forms of the scalar product (Equations 1-8 and 1-12) are equivalent (see Problem 1.4-7).

The definition given in Equation (1-8) for the scalar product of two three-dimensional vectors can be generalized readily for the multiplication of vectors of higher dimension. Assuming now that **A** and **B** are n-dimensional vectors, that is,

$$\mathbf{A} = [A_1 \quad A_2 \quad \ldots \quad A_n] \tag{1-13a}$$

$$\mathbf{B} = [B_1 \quad B_2 \quad \ldots \quad B_n] \tag{1-13b}$$

their scalar product is defined as

$$\mathbf{A} \cdot \mathbf{B} = A_1B_1 + A_2B_2 + \ldots + A_nB_n = \sum_{i=1}^{n} A_iB_i \tag{1-14}$$

As an example, consider again the five-dimensional vectors **A** and **B** given by Equations (1-4a and b). The scalar product of these vectors is

$$\begin{aligned}\mathbf{A} \cdot \mathbf{B} &= [7 \quad -1 \quad 4 \quad 2 \quad -8] \cdot [1 \quad 2 \quad 9 \quad 0 \quad -4] \\ &= 7(1) + (-1)(2) + 4(9) + 2(0) + (-8)(-4) \\ &= 73\end{aligned}$$

The same method of calculation is used if **A** and **B** are column vectors instead of row vectors, or if one is a row vector and the other a column vector. As in the case of addition, it is seen that the vectors must be of the same dimension in order to be conformable for scalar multiplication.

The *magnitude* (or length) of an n-dimensional vector can be found by taking the square root of the scalar product of the vector with itself. In other words, the magnitude of the vector **A** (Equation 1-13a) is

$$A = (\mathbf{A} \cdot \mathbf{A})^{1/2} = (A_1^2 + A_2^2 + \ldots + A_n^2)^{1/2} \tag{1-15}$$

Although this definition applies to vectors of any dimension, it is apparent that in the special case of three dimensions it reduces to the familiar statement that the square of the length of the vector is equal to the sum of the squares of its three scalar components. The term *length* has physical significance only in one, two, or three dimensions; hence, the term *magnitude* seems preferable when discussing vectors of arbitrary dimension.

A *unit vector* is a vector having a magnitude equal to unity, as in the case of the unit vectors **i**, **j**, and **k** in three dimensions. The process of obtaining a unit vector in the same "direction" as any given n-dimensional vector is described in the next section.

An interesting example of scalar multiplication occurs when a vector is multiplied by the *sum vector*, defined as

$$\mathbf{S}_n = [1 \quad 1 \quad \ldots \quad 1] \tag{1-16}$$

in which the subscript n indicates the dimension of the vector. Taking the scalar product of $\mathbf{S}_n$ and the n-dimensional vector **A** (see Equation 1-13a) produces

$$\mathbf{S}_n \cdot \mathbf{A} = A_1 + A_2 + \ldots + A_n = \sum_{i=1}^{n} A_i \tag{1-17}$$

which is the sum of the components of the vector.

The second kind of product obtained from the multiplication of vectors is the *vector product* (also called the *cross product*). For three-dimensional vectors the definition of the vector product **A X B** of the vectors **A** and **B** (see Equations 1-7a and 1-7b) is

$$\mathbf{A} \times \mathbf{B} = (A_y B_z - A_z B_y)\mathbf{i} + (A_z B_x - A_x B_z)\mathbf{j} + (A_x B_y - A_y B_x)\mathbf{k} \tag{1-18a}$$

$$= [A_y B_z - A_z B_y \quad A_z B_x - A_x B_z \quad A_x B_y - A_y B_x] \tag{1-18b}$$

This definition of the vector product can be written in the form of a determinant,* as follows:

---

*The evaluation of determinants is described in Chapter 3.

$$\mathbf{A} \times \mathbf{B} = \begin{vmatrix} \mathbf{i} & \mathbf{j} & \mathbf{k} \\ A_x & A_y & A_z \\ B_x & B_y & B_z \end{vmatrix} \tag{1-18c}$$

It is seen from the above definitions that the vector product is itself a vector quantity. Furthermore, it can be demonstrated (see Problem 1.4-9) that this vector is perpendicular to the plane containing **A** and **B**, and its direction is given by the right-hand rule when rotating from the direction of **A** toward the direction of **B**, as illustrated in Figure 1-5.

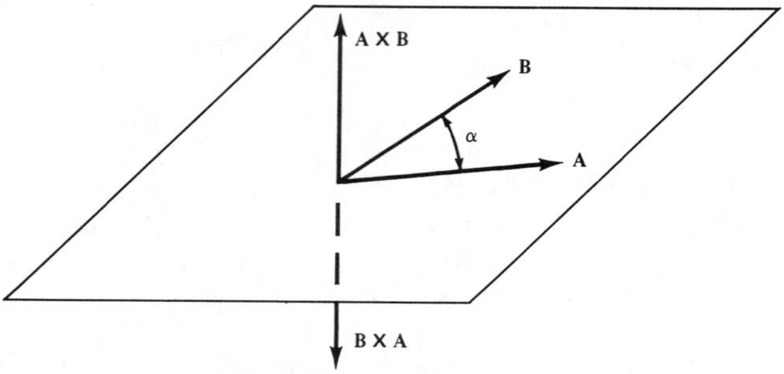

Figure 1-5. Vector product

The vector product is not commutative, and the use of Equations (1-18) in evaluating **B X A** shows that

$$\mathbf{A} \times \mathbf{B} = -\mathbf{B} \times \mathbf{A} \tag{1-19}$$

In other words, the vector **B X A** is the same as **A X B** except that its direction has been reversed, as shown in Figure 1-5.

The scalar value C of the vector product $\mathbf{C} = \mathbf{A} \times \mathbf{B}$ can be given in simplified form by the expression

$$C = AB \sin \alpha \tag{1-20}$$

in which α is the smaller angle between the vectors **A** and **B** (see Figure 1-5). It is left as an exercise (see Problem 1.4-10) to show that the scalar value of the vector product as given by Equation (1-20) is the same as that obtained from any of Equations (1-18). From those equations it is apparent also that the following relations hold for the unit vectors:

$$\mathbf{i} \times \mathbf{j} = \mathbf{k} \qquad \mathbf{j} \times \mathbf{k} = \mathbf{i} \qquad \mathbf{k} \times \mathbf{i} = \mathbf{j} \tag{1-21}$$

Furthermore, the vector product of any vector with itself is a zero vector.

The vector product can be generalized to more than three dimensions, but the generalization has no useful role in matrix algebra so it will not be discussed here.

1.5. Orthogonal and Normalized Vectors. It was pointed out in the preceding section that the scalar product of two three-dimensional vectors is zero if the vectors are orthogonal to each other. The concept of orthogonality can be extended readily to vectors of any dimensions. Thus, two n-dimensional vectors are said to be orthogonal if their scalar product is zero. As an example, consider the six-dimensional vectors **A** and **B** given below:

$$\mathbf{A} = [3 \quad -2 \quad 3 \quad 0 \quad 4 \quad 1] \qquad \mathbf{B} = [-2 \quad 1 \quad 1 \quad -5 \quad 2 \quad -3]$$

The scalar product C of **A** and **B** is zero, as shown:

$$C = \mathbf{A} \cdot \mathbf{B} = 3(-2) - 2(1) + 3(1) + 0(-5) + 4(2) + 1(-3) = 0$$

and therefore the vectors **A** and **B** are orthogonal vectors.

A *normalized* vector is a vector of unit magnitude that is obtained from another vector by dividing every component of that vector by its magnitude. Thus, the normalized vector is a unit vector having the same "direction" (in n-dimensional space) as the given vector. As an example, consider the six-dimensional vector

$$\mathbf{A} = [3 \quad -2 \quad 3 \quad 0 \quad 4 \quad 1]$$

The magnitude of **A** is

$$|\mathbf{A}| = \sqrt{9 + 4 + 9 + 0 + 16 + 1} = \sqrt{39}$$

and therefore the unit vector obtained by normalizing the vector **A** is

$$\mathbf{a} = \frac{1}{\sqrt{39}}\mathbf{A} = \left[\frac{3}{\sqrt{39}} \quad -\frac{2}{\sqrt{39}} \quad \frac{3}{\sqrt{39}} \quad 0 \quad \frac{4}{\sqrt{39}} \quad \frac{1}{\sqrt{39}}\right]$$

The method of normalization illustrated here makes the magnitude of the vector equal to unity, which is the most common kind of normalization. However, other methods are sometimes used; for instance, the vector can be normalized with respect to a particular component, such as the first, last, or largest component.

Problems

1.3-1. Determine the sums **A** + **B** and **C** + **D** and the differences **A** − **B** and **C** − **D** if the vectors **A**, **B**, **C**, and **D** are given as follows:

$$\mathbf{A} = [7.2 \quad -4.3 \quad 0.6 \quad 1.7]$$

$$\mathbf{B} = [-11.0 \quad 11.8 \quad 2.4 \quad -1.9]$$

$$\mathbf{C} = \begin{bmatrix} 1.7 \\ 1.0 \\ -1.0 \\ 4.3 \end{bmatrix} \qquad \mathbf{D} = \begin{bmatrix} -2.4 \\ -0.7 \\ -6.8 \\ 3.0 \end{bmatrix}$$

1.3–2. Obtain the vectors $\mathbf{R}_1$ and $\mathbf{R}_2$ given by the following expressions:

$$\mathbf{R}_1 = 3\mathbf{A} - 2\mathbf{B} \quad \mathbf{R}_2 = 5\mathbf{C} + 2\mathbf{D}$$

if **A**, **B**, **C**, and **D** are the four-dimensional vectors given in the preceding problem.

1.3–3. Find the vector $\mathbf{R}_3$ in order that the following equation is satisfied, assuming that **A** and **B** are the five-dimensional vectors given in Equations (1–4a and b):

$$2\mathbf{A} - 3\mathbf{B} + \mathbf{R}_3 = \mathbf{0}$$

1.3–4. Find the vector $\mathbf{R}_4$ in order that the following equation is satisfied, if **C** and **D** are the four-dimensional vectors given in Problem 1.3–1:

$$\mathbf{R}_4 + 4\mathbf{C} + 3\mathbf{D} = \mathbf{0}$$

1.4–1. Determine the scalar products $\mathbf{A}\cdot\mathbf{B}$ and $\mathbf{C}\cdot\mathbf{D}$ for the vectors given in Problem 1.3–1.

1.4–2. Find the vector **R** according to the expression

$$\mathbf{R} = (\mathbf{A}\cdot\mathbf{B})(2\mathbf{A} + \mathbf{B})$$

if **A** and **B** are the five-dimensional vectors given in Equations (1–4a and b).

1.4–3. Determine the magnitudes A, B, C, and D of the four vectors **A**, **B**, **C**, and **D** in Problem 1.3–1.

1.4–4. Find the component of the vector **A** in the direction of vector **B** if

$$\mathbf{A} = [2 \quad -3 \quad 5] \qquad \mathbf{B} = [1 \quad 4 \quad -2]$$

1.4–5. Determine the angle α between the two vectors **A** and **B** given in the preceding problem.

1.4–6. Solve Problem 1.4–4 if

$$\mathbf{A} = [7 \quad -4 \quad 0 \quad 2] \qquad \mathbf{B} = [-8 \quad 3 \quad 2 \quad -4]$$

1.4–7. Show that the scalar product of two three-dimensional vectors **A** and **B** (see Figure 1–4) as defined by Equation (1–12) is equivalent to that given in Equation (1–8). (Hint: Introduce a vector **C** equal to $\mathbf{A} - \mathbf{B}$, and use the law of cosines.)

1.4–8. Find the vector products $\mathbf{A} \times \mathbf{B}$ and $\mathbf{B} \times \mathbf{A}$ for the following vectors:

$$\mathbf{A} = [6 \quad -3 \quad 2]$$
$$\mathbf{B} = [4 \quad 7 \quad -1]$$

assuming that the given components are in the x, y, and z directions, respectively.

1.4–9. Show that the vector product $\mathbf{A} \times \mathbf{B}$ as defined by Equations (1–18) is orthogonal to the vectors **A** and **B** (and hence is perpendicular to the plane containing **A** and **B**).

1.4–10. Show that the scalar value of the vector product as given by Equation (1–20) agrees with that obtained from Equations (1–18). (Hint: Introduce a vector **C** equal to $\mathbf{A} - \mathbf{B}$; then use the law of cosines and the relation $\sin^2 \alpha + \cos^2 \alpha = 1$.)

1.5–1. Determine the unknown element x in the vector

$$\mathbf{A} = [3.5 \quad -2.0 \quad x \quad 7.0]$$

in order that **A** is orthogonal with the vector

$$\mathbf{B} = [-1.0 \quad 1.5 \quad -0.5 \quad 0.5]$$

1.5–2. Determine the unknown elements x_1 and x_2 in the vector

$$\mathbf{C} = [x_1 \quad x_2 \quad -18]$$

in order that **C** is orthogonal with the vectors

$$\mathbf{D} = [2 \quad 3 \quad -1] \quad \text{and} \quad \mathbf{E} = [3 \quad 0 \quad 1]$$

1.5–3. Solve the preceding problem if

$$\mathbf{C} = [x_1 \quad x_2 \quad -6 \quad 3]$$

$$\mathbf{D} = [3 \quad -2 \quad 1 \quad -4] \qquad \mathbf{E} = [6 \quad 3 \quad 0 \quad 2]$$

1.5–4. Determine the unknown elements x_1, x_2, and x_3 in the following vectors, in order that all three vectors are mutually orthogonal:

$$\mathbf{F} = [-2 \quad 1 \quad x_1 \quad 6] \qquad \mathbf{G} = [3 \quad 0 \quad x_2 \quad 1] \qquad \mathbf{H} = [0 \quad x_3 \quad 1 \quad -4]$$

1.5–5. Obtain the unit vectors **a** and **b** having the same "directions" as the vectors **A** and **B**, respectively, given by Equations (1–4a and b).

1.5–6. Find a three-dimensional unit vector $\mathbf{e}_1$ that is orthogonal to the vectors

$$\mathbf{A} = [2 \quad -1 \quad 3] \qquad \mathbf{B} = [-1 \quad 0 \quad 2]$$

1.5–7. Obtain a unit vector $\mathbf{e}_2$ that is orthogonal to the vectors

$$\mathbf{A} = [-3 \quad 2 \quad 0 \quad 1] \qquad \mathbf{B} = [5 \quad 1 \quad 2 \quad -1] \qquad \mathbf{C} = [-2 \quad 2 \quad 0 \quad 2]$$

2

...Matrices...

2.1. Definitions. In its most common form a matrix is a rectangular array of numbers that can be represented as follows:

$$\mathbf{A} = \begin{bmatrix} A_{11} & A_{12} & \dots & A_{1n} \\ A_{21} & A_{22} & \dots & A_{2n} \\ \dots & \dots & \dots & \dots \\ A_{m1} & A_{m2} & \dots & A_{mn} \end{bmatrix} \tag{2-1}$$

The numbers A_{11}, A_{12}, . . ., A_{mn} That make up the array are called the *elements* of the matrix. The first subscript for the element denotes the *row* and the second denotes the *column* in which the element appears. The square brackets shown in Equation (2-1) will be used to indicate a matrix whenever the elements of the array are shown; otherwise, a single bold face letter (for example, **A**) will be used to represent the matrix. An alternate representation is to enclose within brackets the symbol for the general element of the array; thus, the matrix **A** in Equation (2-1) may also be designated as $[A_{ij}]$ or $[A]$.

The *order* (or *size*) of a matrix is determined by the number of rows and columns; thus, the matrix **A** in Equation (2-1) is of order m by n, usually written as $m \times n$. A *square matrix* has the same number of rows and columns ($m = n$) and is said to be of order n. The elements obtained from a square matrix by beginning at the upper-left corner and proceeding in a diagonal direction to the lower-right corner (that is, the elements A_{11}, A_{22}, . . ., A_{nn}) constitute the *principal diagonal* of the matrix. Similarly, the elements from lower-left to upper-right (A_{n1}, . . ., A_{1n}) constitute the *secondary diagonal*.

A vector can be considered as a special case of a matrix having only one row or one column. A row vector containing n elements is a $1 \times n$ matrix (or *row matrix*), and a column vector of n elements is an $n \times 1$ matrix (*column matrix*). From another point of view, it is useful to introduce the idea of the *dimension* of a matrix, which is the number of subscripts required to locate uniquely an element in the matrix. A rectangular matrix such as the one shown in Equation (2–1) requires two subscripts to locate an element, and hence it is a two-dimensional matrix. A vector, on the other hand, requires only one subscript to locate an element; for example, we can write a row vector in the form

$$\mathbf{A} = [A_1 \quad A_2 \quad \ldots \quad A_n]$$

Therefore, a vector can be considered as a one-dimensional matrix. A three-dimensional matrix of order $m \times n \times p$ consists of p two-dimensional arrays, each of which is of size $m \times n$. This process can be continued indefinitely to matrices of any dimension. However, only one- and two-dimensional matrices will be used in this book.

Up to now the matrix **A** (Equation 2–1) has been discussed as if its elements were numbers. In a more general situation, the elements may be almost any mathematical entities, such as trigonometric functions, algebraic expressions, derivatives, integrals, or even other matrices.

2.2. Operations. The rules for performing mathematical operations (such as addition and multiplication) with matrices are somewhat intuitive and have been formulated in such a way as to make them useful in practical calculations. The simplest relationship involves the *equality* of two matrices. In order that two matrices be equal they must be of the same order and must be identical element by element. Hence, if **A** = **B**, where **A** and **B** are each rectangular matrices of order $m \times n$, then it must be true that $A_{ij} = B_{ij}$ for i from 1 to m and for j from 1 to n.

The *addition* of two matrices **A** and **B** is possible only if the matrices are of the same order. Such matrices are said to be *conformable for addition*. The sum of the two matrices is another matrix of the same order, in which each element is obtained as the sum of the corresponding elements from the original two matrices. In other words, if the sum **A** + **B** = **C** is to be calculated, then each element of **C** must be formed as follows:

$$A_{ij} + B_{ij} = C_{ij}$$

The addition procedure is illustrated by the following example:

$$\mathbf{A} = \begin{bmatrix} 1 & -2 & 3 \\ -1 & 0 & 4 \end{bmatrix} \qquad \mathbf{B} = \begin{bmatrix} 6 & 5 & -2 \\ 0 & 0 & 4 \end{bmatrix}$$

$$\mathbf{A} + \mathbf{B} = \mathbf{C} = \begin{bmatrix} 7 & 3 & 1 \\ -1 & 0 & 8 \end{bmatrix}$$

It is not difficult to see from the manner of performing the addition of matrices that the process is commutative and associative; in other words, **A** + **B** = **B** + **A** and **A** + (**B** + **C**) = (**A** + **B**) + **C**. This fact means that in performing matrix addition the matrices can be arranged in any desired order and grouped in any convenient fashion.

The *subtraction* of matrices is performed in a manner similar to addition and has the same requirement for conformability. As an example, if **A** − **B** = **D**, where **A** and **B** are the matrices given in the previous paragraph, then

$$\mathbf{D} = \begin{bmatrix} -5 & -7 & 5 \\ -1 & 0 & 0 \end{bmatrix}$$

Any matrix can be multiplied by a scalar by multiplying every element of the matrix by the scalar. Hence, if **A** is a matrix of order $m \times n$ and λ is a scalar, the product $\lambda\mathbf{A}$ produces a matrix **B** of order $m \times n$ with $B_{ij} = \lambda A_{ij}$. Using the matrix **D** given above as an example, one may see that the following are true:

$$3\mathbf{D} = \begin{bmatrix} -15 & -21 & 15 \\ -3 & 0 & 0 \end{bmatrix} \qquad -\mathbf{D} = \begin{bmatrix} 5 & 7 & -5 \\ 1 & 0 & 0 \end{bmatrix}$$

This rule for *multiplying a matrix by a scalar* is in agreement with the rules for addition and subtraction, as can be seen by calculating the results given above from the relations 3**D** = **D** + **D** + **D** and −**D** = **D** − **D** − **D**. Also, it is evident that the following properties will hold:

$$(\lambda_1 + \lambda_2)\mathbf{A} = \lambda_1\mathbf{A} + \lambda_2\mathbf{A}$$

$$\lambda(\mathbf{A} + \mathbf{B}) = \lambda\mathbf{A} + \lambda\mathbf{B}$$

$$\lambda_1(\lambda_2\mathbf{A}) = (\lambda_1\lambda_2)\mathbf{A}$$

Thus, when operations involving multiplication by a scalar are to be performed, the matrices can be grouped in a variety of ways.

The next operation to be considered is that of multiplying one matrix by another. This operation, known as matrix multiplication, is described in the next two sections. The subject of matrix inversion, which is analogous to division, is postponed until Chapter 4 in order that some preliminary topics, including determinants, may be discussed first.

2.3. Multiplication of Matrices. There are numerous ways in which two matrices might be multiplied, such as simply multiplying corresponding elements of the matrices. However, such a definition of matrix multiplication would not be useful in practical situations. Since much of the work with matrices is concerned with the solution of simultaneous linear equations and the substitution of new variables by linear transformations, it is natural that matrix multiplication has been defined so as to facilitate these operations.

In order to show the relation of matrix multiplication to the operations just mentioned, consider two simultaneous equations as follows:

$$\begin{aligned} A_{11}X_1 + A_{12}X_2 &= D_1 \\ A_{21}X_1 + A_{22}X_2 &= D_2 \end{aligned} \tag{2-2}$$

in which X_1 and X_2 are considered as unknown quantities and the A's and D's are assumed to be constants. To simplify the writing of these equations, three matrices are now introduced; the first contains the coefficients that appear on the left-hand sides of the equal signs, the second contains the unknowns, and the last contains the terms on the right-hand sides. These matrices are

$$\mathbf{A} = \begin{bmatrix} A_{11} & A_{12} \\ A_{21} & A_{22} \end{bmatrix} \qquad \mathbf{X} = \begin{bmatrix} X_1 \\ X_2 \end{bmatrix} \qquad \mathbf{D} = \begin{bmatrix} D_1 \\ D_2 \end{bmatrix} \tag{2-3}$$

If the process of multiplication is defined in an appropriate manner, it will be possible to write Equations (2-2) in the matrix form

$$\mathbf{AX} = \mathbf{D} \tag{2-4}$$

or

$$\begin{bmatrix} A_{11} & A_{12} \\ A_{21} & A_{22} \end{bmatrix} \begin{bmatrix} X_1 \\ X_2 \end{bmatrix} = \begin{bmatrix} D_1 \\ D_2 \end{bmatrix} \tag{2-5}$$

Comparison of Equations (2-2) and (2-5) shows the rule that must be followed for multiplying the matrices **A** and **X**. If the elements of the first row of the **A** matrix are multiplied by the corresponding elements of the column in the **X** matrix and then added, the result will be the first element in the **D** matrix (in other words, $A_{11}X_1 + A_{12}X_2 = D_1$). Similarly, multiplication of elements from the second row of the **A** matrix with corresponding elements of **X** will give the second element in the **D** matrix (that is, $A_{21}X_1 + A_{22}X_2 = D_2$).

Recalling now the definition of the scalar product of two vectors (see Equation 1-8), it is seen that each element of the **D** matrix is found by taking the scalar product of two vectors, the first of which is the corresponding row vector from the **A** matrix and the second of which is the column vector that constitutes the **X** matrix. However, when multiplying matrices it is common to refer to the process just described as taking the *inner product* of a row and column, instead of using the term *scalar product*.

The rules given above for multiplying a vector by a matrix have practical usefulness when writing simultaneous equations in matrix form. In a more general situation, however, it may be necessary to multiply a matrix (consisting of several vectors) by another matrix. In order to develop a meaningful set of rules for performing this operation, suppose that two new variables Y_1 and Y_2 are related to X_1 and X_2 by the equations

$$\begin{aligned} X_1 &= B_{11}Y_1 + B_{12}Y_2 \\ X_2 &= B_{21}Y_1 + B_{22}Y_2 \end{aligned} \tag{2-6}$$

These new relations can be expressed in the matrix form

$$\mathbf{X} = \mathbf{BY} \tag{2-7}$$

in which

$$\mathbf{B} = \begin{bmatrix} B_{11} & B_{12} \\ B_{21} & B_{22} \end{bmatrix} \qquad \mathbf{Y} = \begin{bmatrix} Y_1 \\ Y_2 \end{bmatrix}$$

and the matrix **X** is the same as in Equations (2-3). Substitution of the expressions for X_1 and X_2 in terms of Y_1 and Y_2 (see Equations 2-6) into the original simultaneous equations constitutes a *change of variables*. (Since the coefficients B in Equations 2-6 are constants, a change of variables of this nature is referred to as a *linear transformation*.) Thus, when X_1 and X_2 from Equations (2-6) are substituted into Equations (2-5), the latter equation becomes

$$\begin{bmatrix} A_{11} & A_{12} \\ A_{21} & A_{22} \end{bmatrix} \begin{bmatrix} B_{11}Y_1 + B_{12}Y_2 \\ B_{21}Y_1 + B_{22}Y_2 \end{bmatrix} = \begin{bmatrix} D_1 \\ D_2 \end{bmatrix} \tag{2-8}$$

which may be stated in more condensed form by substitution of Equation (2-7) into Equation (2-4). Thus,

$$\mathbf{A}(\mathbf{BY}) = \mathbf{D} \tag{2-9}$$

Since the product **BY** in Equation (2-9) is a vector, premultiplication of **BY** by the matrix **A** can be accomplished by the rules stated above for matrix-vector multiplication to yield the following results:

$$\begin{bmatrix} A_{11}(B_{11}Y_1 + B_{12}Y_2) + A_{12}(B_{21}Y_1 + B_{22}Y_2) \\ A_{21}(B_{11}Y_1 + B_{12}Y_2) + A_{22}(B_{21}Y_1 + B_{22}Y_2) \end{bmatrix} = \begin{bmatrix} D_1 \\ D_2 \end{bmatrix}$$

If the coefficients of Y_1 and Y_2 are collected into a 2×2 matrix, this equation can be restated as

$$\begin{bmatrix} (A_{11}B_{11} + A_{12}B_{21}) & (A_{11}B_{12} + A_{12}B_{22}) \\ (A_{21}B_{11} + A_{22}B_{21}) & (A_{21}B_{12} + A_{22}B_{22}) \end{bmatrix} \begin{bmatrix} Y_1 \\ Y_2 \end{bmatrix} = \begin{bmatrix} D_1 \\ D_2 \end{bmatrix} \tag{2-10}$$

or equivalently as

$$\mathbf{CY} = \mathbf{D} \tag{2-11}$$

in which the matrix **C** is the 2×2 array of coefficients in Equation (2-10).

A comparison of Equation (2-11) with Equation (2-9) shows that

$$\mathbf{AB} = \mathbf{C}$$

or

$$\begin{bmatrix} A_{11} & A_{12} \\ A_{21} & A_{22} \end{bmatrix} \begin{bmatrix} B_{11} & B_{12} \\ B_{21} & B_{22} \end{bmatrix} = \begin{bmatrix} C_{11} & C_{12} \\ C_{21} & C_{22} \end{bmatrix}$$
$$= \begin{bmatrix} (A_{11}B_{11} + A_{12}B_{21}) & (A_{11}B_{12} + A_{12}B_{22}) \\ (A_{21}B_{11} + A_{22}B_{21}) & (A_{21}B_{12} + A_{22}B_{22}) \end{bmatrix} \tag{2-12}$$

This equation demonstrates the general rule for multiplying matrices. It is seen that the element C_{11} is obtained as the inner product of the first row of the **A** matrix and the first column of the **B** matrix; the element C_{12} is equal to the inner product of the first row of **A** and the second column of **B**; and so on. Each element of **C** is obtained as the inner product of a row vector from **A** and a column vector from **B**. The particular row in **A** and column in **B** corresponds in each case to the row and column of the element in **C** to be calculated. This example shows also that in order to carry out a matrix multiplication, there must be as many columns of the first matrix as there are rows of the second matrix.

The particular examples used in the above discussion have pointed the way to a meaningful definition of matrix multiplication. In the general case it may be necessary to multiply a matrix **A** of order $m \times n$ and a matrix **B** of order $n \times p$. The product will be a matrix **C** of order $m \times p$, as follows:

$$\mathbf{AB} = \mathbf{C} \tag{2-13}$$

or

$$\begin{bmatrix} A_{11} & A_{12} & \dots & A_{1n} \\ A_{21} & A_{22} & \dots & A_{2n} \\ \dots & \dots & \dots & \dots \\ A_{m1} & A_{m2} & \dots & A_{mn} \end{bmatrix} \begin{bmatrix} B_{11} & B_{12} & \dots & B_{1p} \\ B_{21} & B_{22} & \dots & B_{2p} \\ \dots & \dots & \dots & \dots \\ B_{n1} & B_{n2} & \dots & B_{np} \end{bmatrix} = \begin{bmatrix} C_{11} & C_{12} & \dots & C_{1p} \\ C_{21} & C_{22} & \dots & C_{2p} \\ \dots & \dots & \dots & \dots \\ C_{m1} & C_{m2} & \dots & C_{mp} \end{bmatrix}$$

The multiplication is carried out by taking the inner products of the rows of **A** and the columns of **B**. The element C_{11} in the first row and first column of **C** is obtained as the inner product of the first row of **A** and the first column of **B**, so that

$$C_{11} = A_{11}B_{11} + A_{12}B_{21} + \dots + A_{1n}B_{n1}$$

The element C_{1p} is obtained as the inner product of the first row of **A** and the pth column of **B**, as follows:

$$C_{1p} = A_{11}B_{1p} + A_{12}B_{2p} + \ldots + A_{1n}B_{np}$$

The elements in the second row of the **C** matrix are calculated by taking the inner products of the second row of **A** with the various columns of **B**:

$$\begin{aligned} C_{21} &= A_{21}B_{11} + A_{22}B_{21} + \ldots + A_{2n}B_{n1} \\ C_{22} &= A_{21}B_{12} + A_{22}B_{22} + \ldots + A_{2n}B_{n2} \\ \ldots & \quad \ldots \quad\quad \ldots \quad\quad \ldots \quad \ldots \\ C_{2p} &= A_{21}B_{1p} + A_{22}B_{2p} + \ldots + A_{2n}B_{np} \end{aligned}$$

This method of calculating the elements of the C matrix (or *product matrix*) shows that in order to be *conformable for multiplication* the number of columns of **A** must always be the same as the number of rows of **B**. Except for this requirement, there is no restriction as to the order of these matrices. The matrix **C** will always have as many rows as **A** and as many columns as **B**. These conformability relationships for matrix multiplication can be represented schematically as

$$[m \times n][n \times p] = [m \times p] \tag{2-14}$$

When referring to the product **AB**, the matrix **A** is called the *premultiplier* and **B** the *postmultiplier*. Also, **A** is said to be postmultiplied by **B**, and **B** is said to be premultiplied by **A**.

Matrix multiplication is not commutative in most instances. That is,

$$\mathbf{AB} \neq \mathbf{BA}$$

except in certain special cases that will be mentioned later. It can be seen immediately that if **A** and **B** are rectangular matrices of order $m \times n$ and $n \times p$, they are always conformable for premultiplication of **B** by **A**, but they will be conformable for premultiplication of **A** by **B** only if $p = m$. In such a case the product **AB** will be a square matrix of order $m \times m$, whereas the product **BA** will be a square matrix of order $n \times n$. Hence, the only possible way for the matrices **A** and **B** to be commutative (that is, for **AB** to be equal to **BA**) is for m and n to be the same. Thus, the general conclusion is reached that a necessary condition for **A** and **B** to be commutative (or *permutable*) is that they must be square matrices of the same order. This condition is definitely not a sufficient one, however, and most such square matrices are not commutative, as will be illustrated in the examples to follow.

The associative and distributive laws do hold for matrix multiplication, and the following relationships are valid:

$$(\mathbf{AB})\mathbf{C} = \mathbf{A}(\mathbf{BC})$$

$$\mathbf{A}(\mathbf{B} + \mathbf{C}) = \mathbf{AB} + \mathbf{AC}$$

$$(\mathbf{A} + \mathbf{B})\mathbf{C} = \mathbf{AC} + \mathbf{BC}$$

On the other hand, in addition to being noncommutative, there are other ways in which matrix multiplication does not follow the familiar rules of scalar algebra. For example, if **AB** = **AC**, it does not follow in all cases that **B** = **C**. This fact is illustrated in Examples 6 and 7 of the next section.

2.4. Examples of Matrix Multiplication. Before proceeding to further study, it is important that a thorough familiarity with matrix multiplication be attained. Therefore, several numerical examples are given in this section. In the first two examples each step in the calculations is shown in detail, but in the later examples some of the steps are omitted. A few of the examples illustrate the properties of matrix multiplication that were mentioned previously.

Example 1. Calculate the product **AB** for the matrices **A** and **B** given below.

$$\mathbf{A} = \begin{bmatrix} 4 & -1 & 0 \\ 2 & 5 & -3 \end{bmatrix} \qquad \mathbf{B} = \begin{bmatrix} -3 \\ 3 \\ 1 \end{bmatrix}$$

Multiplication of **A** times **B** produces the following results:

$$\mathbf{AB} = \begin{bmatrix} 4(-3) - 1(3) + 0(1) \\ 2(-3) + 5(3) - 3(1) \end{bmatrix} = \begin{bmatrix} -15 \\ 6 \end{bmatrix}$$

In this example the product **BA** cannot be obtained, since the number of columns of **B** is not equal to the number of rows of **A**.

Example 2. Calculate the product **AB** for the matrices **A** and **B** given below.

$$\mathbf{A} = \begin{bmatrix} 4 & -1 & 0 \\ 2 & 5 & -3 \end{bmatrix} \qquad \mathbf{B} = \begin{bmatrix} 1 & 2 & 4 \\ -2 & -3 & 0 \\ 0 & 6 & 1 \end{bmatrix}$$

The product **AB** is found as follows:

$$\mathbf{AB} = \begin{bmatrix} 4(1) - 1(-2) + 0(0) & 4(2) - 1(-3) + 0(6) & 4(4) - 1(0) + 0(1) \\ 2(1) + 5(-2) - 3(0) & 2(2) + 5(-3) - 3(6) & 2(4) + 5(0) - 3(1) \end{bmatrix}$$

$$= \begin{bmatrix} 6 & 11 & 16 \\ -8 & -29 & 5 \end{bmatrix}$$

Example 3. Calculate the products **AB** and **BA** for the following matrices:

$$\mathbf{A} = \begin{bmatrix} 1 & 0 & 2 \\ -1 & 1 & 3 \end{bmatrix} \qquad \mathbf{B} = \begin{bmatrix} 2 & 1 \\ -2 & 0 \\ 4 & 5 \end{bmatrix}$$

The results are

$$\mathbf{AB} = \begin{bmatrix} 10 & 11 \\ 8 & 14 \end{bmatrix} \qquad \mathbf{BA} = \begin{bmatrix} 1 & 1 & 7 \\ -2 & 0 & -4 \\ -1 & 5 & 23 \end{bmatrix}$$

The matrices **A** and **B** are conformable for multiplication for both products **AB** and **BA**, but the results are evidently quite different.

Example 4. Calculate the products **AB** and **BA** for the following matrices:

$$\mathbf{A} = \begin{bmatrix} 1 & 2 \\ 3 & 4 \end{bmatrix} \qquad \mathbf{B} = \begin{bmatrix} 5 & 6 \\ 7 & 8 \end{bmatrix}$$

The products are

$$\mathbf{AB} = \begin{bmatrix} 19 & 22 \\ 43 & 50 \end{bmatrix} \qquad \mathbf{BA} = \begin{bmatrix} 23 & 34 \\ 31 & 46 \end{bmatrix}$$

This example shows that even though **A** and **B** are square matrices of the same order, they are not commutative for multiplication.

Example 5. Calculate the product **AB**.

$$\mathbf{A} = \begin{bmatrix} 1 & 2 \\ 3 & 6 \end{bmatrix} \qquad \mathbf{B} = \begin{bmatrix} 2 & -4 \\ -1 & 2 \end{bmatrix}$$

The result is

$$\mathbf{AB} = \begin{bmatrix} 0 & 0 \\ 0 & 0 \end{bmatrix}$$

Thus, the product of two matrices can be a matrix having all elements equal to zero, although none of the elements of the matrices **A** and **B** is equal to zero.

Example 6. Calculate the products **AB** and **AC** for the following matrices:

$$\mathbf{A} = \begin{bmatrix} 1 & 0 \\ 2 & 0 \end{bmatrix} \quad \mathbf{B} = \begin{bmatrix} 2 & 3 \\ -4 & 1 \end{bmatrix} \quad \mathbf{C} = \begin{bmatrix} 2 & 3 \\ 1 & 5 \end{bmatrix}$$

The results are

$$\mathbf{AB} = \begin{bmatrix} 2 & 3 \\ 4 & 6 \end{bmatrix} \quad \mathbf{AC} = \begin{bmatrix} 2 & 3 \\ 4 & 6 \end{bmatrix}$$

This example is one in which **AB** = **AC**, although the matrices **B** and **C** are not equal.

Example 7. Calculate the products **AB** and **AC**.

$$\mathbf{A} = \begin{bmatrix} 1 & 1 & 1 \\ 3 & 3 & 3 \end{bmatrix} \quad \mathbf{B} = \begin{bmatrix} -2 & 3 \\ 1 & -5 \\ 4 & 1 \end{bmatrix} \quad \mathbf{C} = \begin{bmatrix} 4 & 1 \\ -4 & -3 \\ 3 & 1 \end{bmatrix}$$

The products are

$$\mathbf{AB} = \begin{bmatrix} 3 & -1 \\ 9 & -3 \end{bmatrix} \quad \mathbf{AC} = \begin{bmatrix} 3 & -1 \\ 9 & -3 \end{bmatrix}$$

This example again shows that the relation **AB** = **AC** does not imply **B** = **C**.

A method for checking matrix multiplication when performing calculations by hand is described later in Section 2.8.

2.5. Special Types of Matrices. There are many special types of matrices that are encountered frequently in engineering analysis. An example is the *transpose* of a matrix, which is a new matrix formed by interchanging corresponding rows and columns of the original matrix; that is, the first row of the original matrix becomes the first column of the transposed matrix, the second row becomes the second column, and so on. If the original matrix **A** is of order $m \times n$, as shown:

$$\mathbf{A} = \begin{bmatrix} A_{11} & A_{12} & \dots & A_{1n} \\ A_{21} & A_{22} & \dots & A_{2n} \\ \dots & \dots & \dots & \dots \\ A_{m1} & A_{m2} & \dots & A_{mn} \end{bmatrix} \qquad (2\text{-}15)$$

then the transposed matrix, denoted as $\mathbf{A}^T$, will be of order $n \times m$:

$$\mathbf{A^T} = \begin{bmatrix} A_{11} & A_{21} & \dots & A_{m1} \\ A_{12} & A_{22} & \dots & A_{m2} \\ \dots & \dots & \dots & \dots \\ A_{1n} & A_{2n} & \dots & A_{mn} \end{bmatrix} \tag{2-16}$$

From this definition of the transpose of a matrix, it is apparent that the transpose of the transpose is the original matrix itself; thus,

$$(\mathbf{A^T})^\mathbf{T} = \mathbf{A} \tag{2-17}$$

Also, the transpose of the sum of two matrices is the sum of the transposes:

$$(\mathbf{A}_1 + \mathbf{A}_2)^\mathbf{T} = \mathbf{A}_1^\mathbf{T} + \mathbf{A}_2^\mathbf{T} \tag{2-18a}$$

This rule may be generalized to the transpose of the sum of any number of matrices as follows:

$$(\mathbf{A}_1 + \mathbf{A}_2 + \dots + \mathbf{A}_n)^\mathbf{T} = \mathbf{A}_1^\mathbf{T} + \mathbf{A}_2^\mathbf{T} + \dots + \mathbf{A}_n^\mathbf{T} \tag{2-18b}$$

Furthermore, the transpose of the product of two matrices is the product of the transposes, but in reverse order:

$$(\mathbf{A}_1\mathbf{A}_2)^\mathbf{T} = \mathbf{A}_2^\mathbf{T}\mathbf{A}_1^\mathbf{T} \tag{2-19a}$$

For a multiple product, the rule becomes

$$(\mathbf{A}_1\mathbf{A}_2 \dots \mathbf{A}_n)^\mathbf{T} = \mathbf{A}_n^\mathbf{T} \dots \mathbf{A}_2^\mathbf{T}\mathbf{A}_1^\mathbf{T} \tag{2-19b}$$

It is left as an exercise (see Problems 2.5-1 and 2.5-2) to demonstrate these rules.

A *null matrix* (or *zero matrix*) is one in which all elements are zero. A matrix of this type was encountered in Example 5 of the preceding section. A null matrix may be either rectangular or square, and it fulfills a purpose in matrix algebra that is analogous to that of zero in scalar algebra or to that of a null vector in vector algebra. A null matrix usually is indicated as **0**; for example, in expressing a set of homogeneous simultaneous equations in matrix form one may write (compare with Equation 2-4)

$$\mathbf{AX} = \mathbf{0}$$

in which **0** represents a column matrix having all elements equal to zero.

A square matrix having all elements equal to zero except those on the principal diagonal is called a *diagonal matrix*. An example of such a matrix is

$$\mathbf{D}_n = \begin{bmatrix} D_{11} & 0 & \dots & 0 \\ 0 & D_{22} & \dots & 0 \\ \dots & \dots & \dots & \dots \\ 0 & 0 & \dots & D_{nn} \end{bmatrix} \tag{2-20}$$

in which the notation $\mathbf{D}_n$ is used to indicate a diagonal matrix of order n, and the elements D_{ij} of the matrix are zero for $i \neq j$. Matrices of this kind are useful in performing certain matrix operations, such as scaling, which is described in the next paragraph and also in the next section. Also, note that multiplication of two diagonal matrices of the same order is commutative, and the product is also a diagonal matrix.

A special type of diagonal matrix is the *scalar matrix*, which has all elements on the principal diagonal equal to the same scalar λ:

$$\mathbf{S}_n = \begin{bmatrix} \lambda & 0 & \dots & 0 \\ 0 & \lambda & \dots & 0 \\ \dots & \dots & \dots & \dots \\ 0 & 0 & \dots & \lambda \end{bmatrix} \tag{2-21}$$

The name of this type of matrix derives from its equivalence to a scalar multiplier of a matrix. For example, if a square matrix $\mathbf{A}$ of order n is either premultiplied or postmultiplied by the scalar matrix $\mathbf{S}_n$, the result is a matrix in which every element equals λ times the corresponding element of $\mathbf{A}$:

$$\mathbf{S}_n\mathbf{A} = \mathbf{A}\mathbf{S}_n = \lambda\mathbf{A} \tag{2-22}$$

which shows that multiplication of a square matrix and a scalar matrix is commutative.

More generally, a rectangular matrix can be multiplied by a scalar by either pre- or postmultiplying with a scalar matrix. Of course, the scalar matrix must be of appropriate order. If a rectangular matrix $\mathbf{B}$ is of order $m \times n$, then the premultiplier must be of order m, whereas the postmultiplier must be of order n, as follows:

$$\mathbf{S}_m\mathbf{B} = \mathbf{B}\mathbf{S}_n = \lambda\mathbf{B} \tag{2-23}$$

The operation of multiplying all elements of a matrix by a scalar is a special type of *scaling*. In the next section, the scaling of individual rows or columns of a matrix is described.

A *unit matrix* is a scalar matrix in which the value of λ is unity, as follows:

$$\mathbf{I}_n = \begin{bmatrix} 1 & 0 & \dots & 0 \\ 0 & 1 & \dots & 0 \\ \dots & \dots & \dots & \dots \\ 0 & 0 & \dots & 1 \end{bmatrix} \tag{2-24}$$

The unit matrix serves somewhat the same purpose in matrix algebra as does the number one (unity) in scalar algebra. It is also called the *identity matrix* because multiplication of a matrix by it will result in the same matrix. Again taking the case of a square matrix **A** of order n, it can be seen that

$$\mathbf{I}_n\mathbf{A} = \mathbf{A}\mathbf{I}_n = \mathbf{A} \tag{2-25}$$

Similarly, for a rectangular matrix **B** of order $m \times n$ the following equalities are valid:

$$\mathbf{I}_m\mathbf{B} = \mathbf{B}\mathbf{I}_n = \mathbf{B} \tag{2-26}$$

Of course, multiplication of an identity matrix by itself results in the same identity matrix.

Other types of square matrices are *symmetric, skew,* and *skew-symmetric* matrices. A symmetric matrix is one in which the element A_{ij} in the ith row and jth column is equal to the element A_{ji} in the jth row and ith column. This relationship between the elements means that the transpose of a symmetric matrix is the same as the original matrix, or $\mathbf{A} = \mathbf{A}^T$ if **A** is a symmetric matrix. A skew matrix has a negative relationship between elements that are not on the principal diagonal, so that $A_{ij} = -A_{ji}$ for $i \neq j$, and the principal diagonal elements may have any value provided that they are not all zero (A_{ii} not all zero).

If all elements on the principal diagonal of a skew matrix are zero, then the matrix is said to be skew-symmetric (or *antimetric*), and the transpose of the matrix is equal to the negative of the original matrix. Thus, a skew-symmetric matrix meets the following conditions: $A_{ij} = -A_{ji}$ for $i \neq j$, $A_{ii} = 0$, and $\mathbf{A} = -\mathbf{A}^T$. Examples of symmetric, skew, and skew-symmetric matrices, respectively, are

$$\begin{bmatrix} 1 & 2 & 3 \\ 2 & 4 & 5 \\ 3 & 5 & 6 \end{bmatrix} \qquad \begin{bmatrix} 1 & 2 & 3 \\ -2 & 4 & -5 \\ -3 & 5 & 6 \end{bmatrix} \qquad \begin{bmatrix} 0 & 2 & 3 \\ -2 & 0 & -5 \\ -3 & 5 & 0 \end{bmatrix}$$

From the preceding definitions it may be observed that all diagonal matrices are both symmetric and skew.

It is possible to show that any square matrix can be expressed as the sum of a symmetric and a skew-symmetric matrix (see Problem 2.5-5) and that the product of a matrix and its transpose is a symmetric matrix (see Problem 2.5-6). Also, the sum of a matrix and its transpose is a symmetric matrix, and their difference is skew-symmetric. Finally, note that the product of two different symmetric matrices is not necessarily a symmetric matrix, although the product of a symmetric matrix and itself is symmetric.

Square matrices having many zero elements are encountered frequently in engineering analysis. Some of these matrices have been given special names, according to the locations of the zero elements. For example, a *lower triangular matrix* has zero elements above and to the right of the principal diagonal; that is, $A_{ij} = 0$ for $i < j$. The appearance of the matrix is as shown:

$$\mathbf{A} = \begin{bmatrix} A_{11} & 0 & 0 & \dots & 0 \\ A_{21} & A_{22} & 0 & \dots & 0 \\ A_{31} & A_{32} & A_{33} & \dots & 0 \\ \dots & \dots & \dots & \dots & \dots \\ A_{n1} & A_{n2} & A_{n3} & \dots & A_{nn} \end{bmatrix} \tag{2-27a}$$

Similarly, an *upper triangular matrix* has $A_{ij} = 0$ for $i > j$, and the matrix has the form

$$\mathbf{A} = \begin{bmatrix} A_{11} & A_{12} & A_{13} & \dots & A_{1n} \\ 0 & A_{22} & A_{23} & \dots & A_{2n} \\ 0 & 0 & A_{33} & \dots & A_{3n} \\ \dots & \dots & \dots & \dots & \dots \\ 0 & 0 & 0 & \dots & A_{nn} \end{bmatrix} \tag{2-27b}$$

A *band matrix* has zero elements everywhere except along a band or strip that runs diagonally through the matrix, usually (but not necessarily) being centered about the principal diagonal. Thus, the following matrix is an example of a band matrix:

$$\mathbf{A} = \begin{bmatrix} A_{11} & A_{12} & 0 & 0 & 0 & 0 \\ A_{21} & A_{22} & A_{23} & 0 & 0 & 0 \\ 0 & A_{32} & A_{33} & A_{34} & 0 & 0 \\ 0 & 0 & A_{43} & A_{44} & A_{45} & 0 \\ 0 & 0 & 0 & A_{54} & A_{55} & A_{56} \\ 0 & 0 & 0 & 0 & A_{65} & A_{66} \end{bmatrix}$$

The band width in the above illustration is three, and the band is centered on the diagonal; hence, this matrix is said to be *tridiagonal*. In general, the width of the band may be any value up to a number that would include the entire matrix within the band.

Powers of matrices will be denoted by exponents in the same manner as for scalars. For example, the square of a matrix $\mathbf{A}$ will be called $\mathbf{A}^2$, and it is equal to $\mathbf{A}$ times $\mathbf{A}$. In a similar way, higher powers are defined as

$$\mathbf{A}^n = \mathbf{AA} \dots \mathbf{A} \tag{2-28}$$

where the matrix **A** appears n times in the product on the right-hand side. It is obvious that the power of a matrix can be obtained in this manner only if the matrix is square and the exponent n is a positive integer. The zero power of a square matrix is defined as the identity matrix of the same order, except when the matrix is null, in which case the zero power is not defined. In addition, it is not difficult to see that

$$\mathbf{A}^m \mathbf{A}^n = \mathbf{A}^n \mathbf{A}^m = \mathbf{A}^{m+n} \tag{2-29}$$

and

$$(\mathbf{A}^n)^{\mathsf{T}} = (\mathbf{A}^{\mathsf{T}})^n \tag{2-30}$$

provided **A** is square and m and n are positive integers or zero. Negative powers of a matrix are discussed in Section 4.5.

Square Roots of a Matrix. A square root of a square matrix is another square matrix of the same order that will produce the original matrix when multiplied by itself. Hence, **B** is a square root of **A** if

$$\mathbf{B}\,\mathbf{B} = \mathbf{B}^2 = \mathbf{A} \tag{2-31}$$

Square roots of matrices are identified in the customary manner:

$$\mathbf{B} = \sqrt{\mathbf{A}} = \mathbf{A}^{1/2} \tag{2-32}$$

The square root of a matrix is not unique, and most matrices have several different square roots; some even have an infinite number of square roots, as illustrated later. Nevertheless, it is not easy to determine the square root of a matrix except in a few special cases.

The identity matrix (see Equation 2-24) is clearly a special case, and it can be seen immediately that both **I** and −**I** are square roots of **I**. Furthermore, the "ones" on the principal diagonal of **I** can be given plus or minus signs in any desired arrangement, and the matrix will still be a square root of **I**. If a matrix is constructed with "ones" on the secondary diagonal and zeroes elsewhere, then that matrix is also a square root of **I**.

The 2 × 2 identity matrix is seen to have the following square root:

$$\mathbf{B} = \sqrt{\mathbf{I}} = \frac{1}{b}\begin{bmatrix} ab & b^2 \\ 1-a^2 & -ab \end{bmatrix} \tag{2-33}$$

in which a and b can have any values (except that b must be nonzero). For instance, if $a = 2$ and $b = 3$, the square root becomes

$$\mathbf{B} = \sqrt{\mathbf{I}} = \begin{bmatrix} 2 & 3 \\ -1 & -2 \end{bmatrix}$$

From this illustration it is apparent that the 2×2 identity matrix has an infinite number of square roots, both symmetric and unsymmetric. In a similar way, it can be shown that identity matrices of higher order have an infinite number of square roots of great variety.

There are also many possibilities for the square roots of a diagonal matrix (see Equation 2-20). The simplest square root is another diagonal matrix formed by taking square roots of the diagonal terms:

$$\sqrt{\mathbf{D}_n} = \begin{bmatrix} \sqrt{D_{11}} & 0 & \cdots & 0 \\ 0 & \sqrt{D_{22}} & \cdots & 0 \\ \cdots & \cdots & \cdots & \cdots \\ 0 & 0 & \cdots & \sqrt{D_{nn}} \end{bmatrix} \qquad (2\text{-}34)$$

Because each square root on the diagonal can be taken as plus or minus, there are 2^n such diagonal square root matrices. In general, Equation (2-34) gives all of the square roots of a diagonal matrix; however, other forms of square roots exist in certain special cases. For instance, if all diagonal elements are the same, the diagonal matrix will be the identity matrix multiplied by a scalar; hence, it will have an infinite number of square roots.

When a matrix is not diagonal, it is usually difficult to determine any square roots by inspection. Instead, a general method for finding square roots from eigenvalues and eigenvectors must be used; this method is described in Section 6.8.

2.6. Elementary Transformations. There are certain operations, known as *elementary transformations*, that sometimes must be performed on a matrix. These transformations are carried out by means of matrix multiplications. They may be performed on either the rows or the columns of a matrix and are of three kinds: scaling, interchanging, and combining. *Scaling* is the multiplication of all elements in any row (or column) of a matrix by a scalar. Scaling the rows of a matrix is accomplished by premultiplying by an appropriate diagonal matrix; scaling columns is accomplished by postmultiplication. To illustrate the operation of scaling, consider the following example involving the square matrices **E** and **A** of the same order:

$$\mathbf{E} = \begin{bmatrix} 1 & 0 & 0 \\ 0 & \lambda & 0 \\ 0 & 0 & 1 \end{bmatrix} \qquad \mathbf{A} = \begin{bmatrix} A_{11} & A_{12} & A_{13} \\ A_{21} & A_{22} & A_{23} \\ A_{31} & A_{32} & A_{33} \end{bmatrix}$$

Note that **E** is a diagonal matrix with all but one element on the principal diagonal equal to unity. When the products **EA** and **AE** are calculated, the results obtained are

$$\mathbf{EA} = \begin{bmatrix} A_{11} & A_{12} & A_{13} \\ \lambda A_{21} & \lambda A_{22} & \lambda A_{23} \\ A_{31} & A_{32} & A_{33} \end{bmatrix} \qquad \mathbf{AE} = \begin{bmatrix} A_{11} & \lambda A_{12} & A_{13} \\ A_{21} & \lambda A_{22} & A_{23} \\ A_{31} & \lambda A_{32} & A_{33} \end{bmatrix}$$

Inspection of these results shows that the product matrix **EA** differs from **A** only in that all elements of the second row have been multiplied by λ; similarly, in the product **AE**, all elements in the second column of **A** are multiplied by λ.

The matrix **E** in the above example is called an *elementary matrix* and can be obtained by modifying the identity matrix **I** in the same way that the matrix **A** is to be modified. In other words, if it is desired to multiply all elements of the second row of **A** by λ, the required premultiplier is a matrix obtained from the identity matrix by multiplying the second row by λ. A similar statement can be made pertaining to the elementary matrix used as a postmultiplier when it is desired to multiply all elements in the second column by λ.

The results illustrated in the above examples can now be stated more completely for the case of a square matrix. To multiply all elements of any row (or column) of a square matrix by a scalar, it is only necessary to premultiply (or postmultiply) the matrix by an elementary matrix of the same order, provided that the elementary matrix is obtained from the identity matrix by performing the same row (or column) multiplication on it.

Scaling of the rows or columns of a rectangular matrix **B** is performed in a manner similar to that for a square matrix. If the transformation is being performed on rows, then the elementary matrix (which always must be a square matrix) is a premultiplier and must have the same order as the number of rows of **B**. If the transformation is on the columns, then the elementary matrix is a postmultiplier and must have the same order as the number of columns of **B**.

The transformation known as *interchanging* consists of interchanging any two rows (or columns) of a matrix. As in the case of scaling, this operation is accomplished by premultiplying (or postmultiplying) by an elementary matrix. The elementary matrix is formed from the identity matrix by making the desired interchange on it. This idea is readily seen from the following illustrations.

Assume that it is desired to interchange the first and second rows of the matrix

$$\mathbf{A} = \begin{bmatrix} A_{11} & A_{12} & A_{13} \\ A_{21} & A_{22} & A_{23} \\ A_{31} & A_{32} & A_{33} \end{bmatrix}$$

Then the required premultiplier is the elementary matrix

$$\mathbf{E} = \begin{bmatrix} 0 & 1 & 0 \\ 1 & 0 & 0 \\ 0 & 0 & 1 \end{bmatrix}$$

which is obtained from the identity matrix by interchanging the first and second rows. The product **EA** gives

$$\mathbf{EA} = \begin{bmatrix} A_{21} & A_{22} & A_{23} \\ A_{11} & A_{12} & A_{13} \\ A_{31} & A_{32} & A_{33} \end{bmatrix}$$

and when this result is compared with **A** it is seen that the first and second rows have been interchanged.

Next assume that the first and third columns of **A** are to be interchanged. Performing this interchange on the identity matrix gives

$$\mathbf{E} = \begin{bmatrix} 0 & 0 & 1 \\ 0 & 1 & 0 \\ 1 & 0 & 0 \end{bmatrix}$$

and when **A** is postmultiplied by **E**, the result is

$$\mathbf{AE} = \begin{bmatrix} A_{13} & A_{12} & A_{11} \\ A_{23} & A_{22} & A_{21} \\ A_{33} & A_{32} & A_{31} \end{bmatrix}$$

as desired. Similar operations of interchanging can be performed with rectangular matrices.

Finally, the *combining* of various rows (or columns) of a matrix will be considered. Combining means adding a scalar multiple of one row (or column) of the matrix to a second row (or column). In this process, the former row (or column) remains unchanged while the latter row (or column) takes on the new values. Again using as an example the matrix **A** given above, assume that it is necessary to add twice the second row to the third row. When this same transformation is performed on the identity matrix, the result is

$$\mathbf{E} = \begin{bmatrix} 1 & 0 & 0 \\ 0 & 1 & 0 \\ 0 & 2 & 1 \end{bmatrix}$$

Now, if **A** is premultiplied by **E**, the product is

$$\mathbf{EA} = \begin{bmatrix} A_{11} & A_{12} & A_{13} \\ A_{21} & A_{22} & A_{23} \\ 2A_{21} + A_{31} & 2A_{22} + A_{32} & 2A_{23} + A_{33} \end{bmatrix}$$

and the desired transformation is obtained. Similarly, to subtract three times the first column from the third column, it is necessary to postmultiply by the matrix

$$\mathbf{E} = \begin{bmatrix} 1 & 0 & -3 \\ 0 & 1 & 0 \\ 0 & 0 & 1 \end{bmatrix}$$

and thereby obtain

$$\mathbf{AE} = \begin{bmatrix} A_{11} & A_{12} & -3A_{11} + A_{13} \\ A_{21} & A_{22} & -3A_{21} + A_{23} \\ A_{31} & A_{32} & -3A_{31} + A_{33} \end{bmatrix}$$

As with scaling and interchanging, the operation of combining also can be performed with rectangular matrices.

Several elementary transformations can be carried out consecutively by forming multiple products involving elementary matrices. For example, performing the three transformations of (1) multiplying the second row of **A** by three, (2) adding the new second row to the third row, and (3) interchanging the second and third columns, requires a matrix multiplication of the form $\mathbf{E}_2\mathbf{E}_1\mathbf{AE}_3$, in which

$$\mathbf{E}_1 = \begin{bmatrix} 1 & 0 & 0 \\ 0 & 3 & 0 \\ 0 & 0 & 1 \end{bmatrix} \qquad \mathbf{E}_2 = \begin{bmatrix} 1 & 0 & 0 \\ 0 & 1 & 0 \\ 0 & 1 & 1 \end{bmatrix} \qquad \mathbf{E}_3 = \begin{bmatrix} 1 & 0 & 0 \\ 0 & 0 & 1 \\ 0 & 1 & 0 \end{bmatrix}$$

It should be noted that $\mathbf{E}_1$, $\mathbf{E}_2$, and $\mathbf{E}_3$ are the elementary matrices corresponding to the three transformations stated above. The result of the multiplication is

$$\mathbf{E}_2\mathbf{E}_1\mathbf{AE}_3 = \begin{bmatrix} A_{11} & A_{13} & A_{12} \\ 3A_{21} & 3A_{23} & 3A_{22} \\ 3A_{21} + A_{31} & 3A_{23} + A_{33} & 3A_{22} + A_{32} \end{bmatrix}$$

and all three transformations have been accomplished.

In the preceding example, it is essential that the transformations be carried out in the appropriate sequence. The results would be quite different if, for example, the second row were added to the third row before multiplication of the second row by three. In such a case, the sequence of transformations would be as follows: (1) add the second row to the third row; (2) multiply the second row by three; and (3) interchange the second and third columns. This new sequence of transformations is accomplished by the multiplication $\mathbf{E}_1\mathbf{E}_2\mathbf{AE}_3$, and the result is

$$\mathbf{E}_1\mathbf{E}_2\mathbf{A}\mathbf{E}_3 = \begin{bmatrix} A_{11} & A_{13} & A_{12} \\ 3A_{21} & 3A_{23} & 3A_{22} \\ A_{21} + A_{31} & A_{23} + A_{33} & A_{22} + A_{32} \end{bmatrix}$$

From these two examples, it is seen that the sequence from right to left of the elementary matrices used as premultipliers of **A** must correspond to the sequence of the desired row transformations. Also, the sequence from left to right of elementary matrices used as postmultipliers must correspond to the desired sequence of column transformations.

Instead of carrying out the successive transformations by means of several elementary matrices, as in the preceding examples, it is possible to construct one *transformation matrix* to serve as a premultiplier and one to serve as a postmultiplier. The transformation matrix for premultiplication will be the same as the product of the elementary matrices used as premultipliers in carrying out the same transformations, and similarly for the postmultiplier. As an example, the transformation $\mathbf{E}_2\mathbf{E}_1\mathbf{A}\mathbf{E}_3$ given previously can be performed by the multiplication $\mathbf{T}_1\mathbf{A}\mathbf{E}_3$, in which $\mathbf{T}_1$ is a transformation matrix given by

$$\mathbf{T}_1 = \mathbf{E}_2\mathbf{E}_1 = \begin{bmatrix} 1 & 0 & 0 \\ 0 & 3 & 0 \\ 0 & 3 & 1 \end{bmatrix}$$

The matrix $\mathbf{T}_1$ can be constructed either by the multiplication of elementary matrices, as shown above, or by beginning with the identity matrix and performing on it the desired row transformations. In so doing, the transformations must be carried out on the identity matrix in the same sequence that they are to be performed on the matrix **A**.

Now using the second example given above, assume that the transformations given by $\mathbf{E}_1\mathbf{E}_2\mathbf{A}\mathbf{E}_3$ are to be performed. Then the premultiplier is

$$\mathbf{T}_2 = \mathbf{E}_1\mathbf{E}_2 = \begin{bmatrix} 1 & 0 & 0 \\ 0 & 3 & 0 \\ 0 & 1 & 1 \end{bmatrix}$$

Again it should be noted that the transformation matrix can be obtained either by multiplying elementary matrices or by operating on the identity matrix with the desired transformations. All of the above conclusions apply in an analogous manner to operations on columns through postmultiplication.

To illustrate the process of performing in one step several transformations, consider the following example. Assume that it is desired to transform the third-order matrix **A** used in the preceding examples in such a manner that (1) the first row is multiplied by four; (2) the second row is added to the first row;

(3) the first and third rows are interchanged; (4) the second column is multiplied by two; and (5) the second column is interchanged with the first column. The row operations can be carried out by a premultiplier $\mathbf{T}_1$ obtained by performing steps (1), (2), and (3) on the identity matrix:

$$\mathbf{T}_1 = \begin{bmatrix} 0 & 0 & 1 \\ 0 & 1 & 0 \\ 4 & 1 & 0 \end{bmatrix}$$

Similarly, the postmultiplier $\mathbf{T}_2$ is found by performing steps (4) and (5) on the identity matrix:

$$\mathbf{T}_2 = \begin{bmatrix} 0 & 1 & 0 \\ 2 & 0 & 0 \\ 0 & 0 & 1 \end{bmatrix}$$

As the last step, the product $\mathbf{T}_1\mathbf{A}\mathbf{T}_2$ is found:

$$\mathbf{T}_1\mathbf{A}\mathbf{T}_2 = \begin{bmatrix} 2A_{32} & A_{31} & A_{33} \\ 2A_{22} & A_{21} & A_{23} \\ 8A_{12} + 2A_{22} & 4A_{11} + A_{21} & 4A_{13} + A_{23} \end{bmatrix}$$

and it is seen that all five of the elementary transformations have been performed by means of the transformation matrices. In summary, a transformation matrix of the kind used in these examples is equal to the product of two or more elementary matrices and can be constructed directly from the identity matrix by performing successively the appropriate elementary transformations. Such transformation matrices will be used later in Section 4.7 for the purpose of calculating the inverse of a matrix.

2.7. Partitioning of Matrices. It is frequently necessary to deal separately with various groups of elements, or *submatrices*, within a larger matrix. This situation can arise when the size of a matrix becomes too large for convenient handling, and it becomes necessary to work with only a portion of the matrix at any one time. Also, there will be cases in which one part of a matrix will have a physical significance that is different from the remainder, and it is instructive to isolate that portion and identify it by a special symbol. Still another example is given in the next section, where an extra row or column is added to a matrix in order to make a computational check on the results of a matrix multiplication.

In all such situations in which a matrix is subdivided into blocks of elements, the matrix is said to be *partitioned*, and dashed lines are used to indicate how the partitioning is accomplished. These lines are shown in the following example, in which the 4×5 matrix $\mathbf{A}$ has been partitioned into four blocks of elements, each of which is itself a matrix:

$$\mathbf{A} = \left[\begin{array}{ccc|cc} A_{11} & A_{12} & A_{13} & A_{14} & A_{15} \\ A_{21} & A_{22} & A_{23} & A_{24} & A_{25} \\ A_{31} & A_{32} & A_{33} & A_{34} & A_{35} \\ \hline A_{41} & A_{42} & A_{43} & A_{44} & A_{45} \end{array}\right]$$

The partitioning lines must always extend entirely through the matrix as in the above example. If the submatrices of **A** are denoted by the symbols $\mathbf{A}_{11}$, $\mathbf{A}_{12}$, $\mathbf{A}_{21}$, and $\mathbf{A}_{22}$ so that

$$\mathbf{A}_{11} = \begin{bmatrix} A_{11} & A_{12} & A_{13} \\ A_{21} & A_{22} & A_{23} \\ A_{31} & A_{32} & A_{33} \end{bmatrix} \qquad \mathbf{A}_{12} = \begin{bmatrix} A_{14} & A_{15} \\ A_{24} & A_{25} \\ A_{34} & A_{35} \end{bmatrix}$$

$$\mathbf{A}_{21} = [A_{41} \quad A_{42} \quad A_{43}] \qquad \mathbf{A}_{22} = [A_{44} \quad A_{45}]$$

then the original matrix can be written in the form

$$\mathbf{A} = \begin{bmatrix} \mathbf{A}_{11} & \mathbf{A}_{12} \\ \mathbf{A}_{21} & \mathbf{A}_{22} \end{bmatrix} \tag{2-35}$$

A partitioned matrix may be transposed by appropriate transposition and rearrangement of the submatrices. For instance, it can be seen by inspection that the transpose of the matrix **A** (Equation 2-35) is

$$\mathbf{A}^{\mathsf{T}} = \begin{bmatrix} \mathbf{A}_{11}^{\mathsf{T}} & \mathbf{A}_{21}^{\mathsf{T}} \\ \mathbf{A}_{12}^{\mathsf{T}} & \mathbf{A}_{22}^{\mathsf{T}} \end{bmatrix} \tag{2-36}$$

Note that $\mathbf{A}^{\mathsf{T}}$ has been formed by transposing each submatrix of **A** and then interchanging the submatrices on the secondary diagonal.

Partitioned matrices such as the one given in Equation (2-35) can be added, subtracted, and multiplied provided that the partitioning is performed in an appropriate manner. For the addition or subtraction of two matrices, it is necessary that both matrices be partitioned in exactly the same way. Thus, a partitioned matrix **B** of order 4×5 (compare with matrix **A** above) will be comformable for addition with **A** only if it is partitioned as follows:

$$\mathbf{B} = \left[\begin{array}{ccc|cc} B_{11} & B_{12} & B_{13} & B_{14} & B_{15} \\ B_{21} & B_{22} & B_{23} & B_{24} & B_{25} \\ B_{31} & B_{32} & B_{33} & B_{34} & B_{35} \\ \hline B_{41} & B_{42} & B_{43} & B_{44} & B_{45} \end{array}\right]$$

This matrix can be expressed in the form

$$\mathbf{B} = \begin{bmatrix} \mathbf{B}_{11} & \mathbf{B}_{12} \\ \mathbf{B}_{21} & \mathbf{B}_{22} \end{bmatrix}$$

in which $\mathbf{B}_{11}$, $\mathbf{B}_{12}$, $\mathbf{B}_{21}$, and $\mathbf{B}_{22}$ represent the corresponding submatrices. In order to add **A** and **B** and obtain a sum **C**, it is necessary according to the rules for addition of matrices that the following represent the sum:

$$\mathbf{A}+\mathbf{B}=\begin{bmatrix}\mathbf{A}_{11}+\mathbf{B}_{11} & \mathbf{A}_{12}+\mathbf{B}_{12}\\ \mathbf{A}_{21}+\mathbf{B}_{21} & \mathbf{A}_{22}+\mathbf{B}_{22}\end{bmatrix}=\begin{bmatrix}\mathbf{C}_{11} & \mathbf{C}_{12}\\ \mathbf{C}_{21} & \mathbf{C}_{22}\end{bmatrix}=\mathbf{C}$$

This result requires that the submatrices $\mathbf{A}_{11}$ and $\mathbf{B}_{11}$, $\mathbf{A}_{12}$ and $\mathbf{B}_{12}$, and so on, be conformable for addition; hence the conclusion is reached that the matrices **A** and **B** must be partitioned identically. Furthermore, the sum matrix **C** also will have the same partitioning.

The conformability requirement for multiplication of partitioned matrices is somewhat different from that for addition and subtraction. To show the requirement, consider again the matrix **A** given previously and assume that it is to be postmultiplied by a matrix **D**, which must have five rows but may have any number of columns. Also, assume that **D** is partitioned into four submatrices as follows:

$$\mathbf{D}=\begin{bmatrix}\mathbf{D}_{11} & \mathbf{D}_{12}\\ \mathbf{D}_{21} & \mathbf{D}_{22}\end{bmatrix}$$

Then, when forming the product **AD** according to the usual rules for matrix multiplication, the following result is obtained:

$$\mathbf{R}=\mathbf{AD}=\begin{bmatrix}\mathbf{A}_{11} & \mathbf{A}_{12}\\ \mathbf{A}_{21} & \mathbf{A}_{22}\end{bmatrix}\begin{bmatrix}\mathbf{D}_{11} & \mathbf{D}_{12}\\ \mathbf{D}_{21} & \mathbf{D}_{22}\end{bmatrix}$$

$$=\begin{bmatrix}\mathbf{A}_{11}\mathbf{D}_{11}+\mathbf{A}_{12}\mathbf{D}_{21} & \mathbf{A}_{11}\mathbf{D}_{12}+\mathbf{A}_{12}\mathbf{D}_{22}\\ \mathbf{A}_{21}\mathbf{D}_{11}+\mathbf{A}_{22}\mathbf{D}_{21} & \mathbf{A}_{21}\mathbf{D}_{12}+\mathbf{A}_{22}\mathbf{D}_{22}\end{bmatrix}$$

$$=\begin{bmatrix}\mathbf{R}_{11} & \mathbf{R}_{12}\\ \mathbf{R}_{21} & \mathbf{R}_{22}\end{bmatrix}$$

In order to determine the submatrices $\mathbf{R}_{11}$, $\mathbf{R}_{12}$, and so on, in the final matrix **R**, it is necessary to form the sum of matrix products: for example, $\mathbf{R}_{11}$ is equal to $\mathbf{A}_{11}\mathbf{D}_{11} + \mathbf{A}_{12}\mathbf{D}_{21}$. These matrix products require that certain conditions must be satisfied. For instance, the product $\mathbf{A}_{11}\mathbf{D}_{11}$ requires that the number of columns of $\mathbf{A}_{11}$ be the same as the number of rows of $\mathbf{D}_{11}$; the product $\mathbf{A}_{12}\mathbf{D}_{21}$ requires that the number of columns of $\mathbf{A}_{12}$ be the same as the number of rows of $\mathbf{D}_{21}$, and so on for the remaining products. All of these conditions will be satisfied if the columns of **A** are partitioned in exactly the same way that the rows of **D** are partitioned. It does not matter how the rows of **A** and the columns of **D** are partitioned.

To illustrate this conclusion, assume that the matrix **D** is of order 5 × 3:

$$\mathbf{D} = \left[\begin{array}{c:cc} D_{11} & D_{12} & D_{13} \\ D_{21} & D_{22} & D_{23} \\ D_{31} & D_{32} & D_{33} \\ \hdashline D_{41} & D_{42} & D_{43} \\ D_{51} & D_{52} & D_{53} \end{array}\right]$$

Then in order to be conformable for premultiplication by **A** it must be partitioned between the third and fourth rows (as shown), inasmuch as **A** is partitioned between the third and fourth columns. The column partitioning of **D** is immaterial and may be between the first and second columns (as shown), between the second and third columns, or there may be no column partitioning at all. The product matrix **AD** will be partitioned in the same way as the rows of **A** and the columns of **D**.

In summary, the multiplication of two matrices that are partitioned in a conformable manner may be accomplished by applying the multiplication rules to the submatrices. By this process the submatrices of the product matrix are obtained. Of course, the final result is the same as if the original unpartitioned matrices had been multiplied directly.

2.8. Checking Matrix Multiplication. The process of matrix multiplication was described earlier in Sections 2.3 and 2.4. When the multiplication is performed by hand, it is useful to have a simple procedure for checking the validity of the results. In this section, a method is described that makes use of partitioning. The method is based upon summing the elements in various rows and columns of the matrices involved in the multiplication.

To show the basis of the checking procedure, consider the following numerical example in which the matrices **A** and **B** are multiplied to give a product **C**:

$$\mathbf{AB} = \mathbf{C} \tag{2-37}$$

or

$$\begin{bmatrix} 1 & 2 & 4 \\ 4 & 0 & 2 \\ 3 & 1 & 1 \end{bmatrix} \begin{bmatrix} 2 & 1 \\ 0 & -1 \\ 1 & 3 \end{bmatrix} = \begin{bmatrix} 6 & 11 \\ 10 & 10 \\ 7 & 5 \end{bmatrix} \tag{2-38}$$

In devising a method for checking the above numerical result for the matrix **C**, it is important to observe the consequences of premultiplying a matrix by a row vector $\mathbf{V}_1$ having each element equal to unity. It is, of course, necessary that $\mathbf{V}_1$ have as many elements as there are rows in the matrix to be premultiplied, as illustrated in the following example:

$$\mathbf{V}_1\mathbf{A} = [1 \quad 1 \quad 1]\begin{bmatrix} 1 & 2 & 4 \\ 4 & 0 & 2 \\ 3 & 1 & 1 \end{bmatrix} = [8 \quad 3 \quad 7]$$

The product $\mathbf{V}_1\mathbf{A}$ is a row matrix in which the elements are the sums of the elements in the columns of **A**. An analogous situation exists if a matrix is postmultiplied by a column vector $\mathbf{V}_2$ having each element equal to unity; thus,

$$\mathbf{BV}_2 = \begin{bmatrix} 2 & 1 \\ 0 & -1 \\ 1 & 3 \end{bmatrix}\begin{bmatrix} 1 \\ 1 \end{bmatrix} = \begin{bmatrix} 3 \\ -1 \\ 4 \end{bmatrix}$$

This product is a column matrix consisting of the sums of the rows of **B**

If both sides of Equation (2-37) are premultiplied by $\mathbf{V}_1$, the equation becomes

$$(\mathbf{V}_1\mathbf{A})\mathbf{B} = \mathbf{V}_1\mathbf{C}$$

This equation shows that when **B** is premultiplied by a row matrix consisting of the column sums of **A**, the result is a row matrix consisting of the column sums of **C**; thus,

$$[8 \quad 3 \quad 7]\begin{bmatrix} 2 & 1 \\ 0 & -1 \\ 1 & 3 \end{bmatrix} = [23 \quad 26] \qquad (2\text{-}39)$$

If both sides of Equation (2-37) are postmultiplied by $\mathbf{V}_2$, the result is

$$\mathbf{A}(\mathbf{BV}_2) = \mathbf{CV}_2$$

This equation states that when **A** is postmultiplied by a column matrix consisting of the row sums of **B**, the result is a column matrix composed of the row sums of **C**; for example, in this problem the equation is

$$\begin{bmatrix} 1 & 2 & 4 \\ 4 & 0 & 2 \\ 3 & 1 & 1 \end{bmatrix}\begin{bmatrix} 3 \\ -1 \\ 4 \end{bmatrix} = \begin{bmatrix} 17 \\ 20 \\ 12 \end{bmatrix} \qquad (2\text{-}40)$$

The column and row sums of the matrix **C** as originally calculated (see Equation 2-38) can be compared with the sums calculated in Equations (2-39) and (2-40), respectively, to give a numerical check on the accuracy.

If desired, one additional check can be made. Premultiplying both sides of Equation (2-37) by $\mathbf{V}_1$ and postmultiplying by $\mathbf{V}_2$ gives the equation

$$(\mathbf{V}_1\mathbf{A})(\mathbf{B}\mathbf{V}_2) = (\mathbf{V}_1\mathbf{C})\mathbf{V}_2 \tag{2–41}$$

In this equation, $\mathbf{V}_1\mathbf{A}$ is a row vector of column sums of **A**, and $\mathbf{B}\mathbf{V}_2$ is a column vector of row sums of **B**. Their product is the sum of all elements in **C**. This conclusion can be seen by examining the right-hand side of the equation, in which the product $\mathbf{V}_1\mathbf{C}$ is a row vector of column sums of **C**. When this product is postmultiplied by $\mathbf{V}_2$, the result is the sum of all elements in $\mathbf{V}_1\mathbf{C}$; that is, the sum of all elements in the product matrix **C**. These observations are illustrated in the following calculation for Equation (2–41):

$$[8 \quad 3 \quad 7]\begin{bmatrix} 3 \\ -1 \\ 4 \end{bmatrix} = [49]$$

The last result in this calculation can also be used as a check since it must equal the sum of all elements in **C**, the sum of all elements in $\mathbf{V}_1\mathbf{C}$ (see Equation 2–39), and the sum of all elements in $\mathbf{C}\mathbf{V}_2$ (see Equation 2–40).

When making calculations by hand, it is desirable that the check procedures be incorporated directly into the original matrix multiplication instead of being performed separately as was done in the illustrations given above. The desired results can be accomplished readily by adding a row to **A** and a column to **B**, the additional row and column being partitioned in each case from the remainder of the matrix. The row and column consist of the column sums of **A** and row sums of **B**, respectively, as shown in the matrix multiplication below:

$$\left[\begin{array}{ccc} 1 & 2 & 4 \\ 4 & 0 & 2 \\ 3 & 1 & 1 \\ \hline 8 & 3 & 7 \end{array}\right]\left[\begin{array}{cc|c} 2 & 1 & 3 \\ 0 & -1 & -1 \\ 1 & 3 & 4 \end{array}\right] = \left[\begin{array}{cc|c} 6 & 11 & 17 \\ 10 & 10 & 20 \\ 7 & 5 & 12 \\ \hline 23 & 26 & 49 \end{array}\right] \tag{2–42}$$

As required, the product matrix on the right-hand side is partitioned according to the rows of the premultiplier and the columns of the postmultiplier. The additional row and column matrices in **C** give a check on the multiplication, since they must represent the row and column sums of **C**; in addition, the submatrix of **C** consisting of one element must be such that this element is equal to the sums of the elements in each of the other three submatrices in **C**.

The operations carried out in Equation (2–42) can now be generalized and the checking procedure stated as follows: (1) add a row to the matrix **A**, producing a new matrix that is partitioned into submatrices **A** and $\mathbf{A}_1$, in which $\mathbf{A}_1$ is obtained by summing the columns in **A** (see Equation 2–43 below); (2) add a column to the matrix **B**, giving a submatrix $\mathbf{B}_1$ that is obtained by summing the rows of **B**; (3) perform the matrix multiplication of the two new matrices, either in the standard way or by using partitioning; (4) partition the resulting product matrix so that $\mathbf{C}_1$ is a row matrix and $\mathbf{C}_2$ is a column matrix, as shown:

$$\begin{bmatrix} \mathbf{A} \\ \hline \mathbf{A}_1 \end{bmatrix} [\mathbf{B} \mid \mathbf{B}_1] = \left[\begin{array}{c|c} \mathbf{C} & \mathbf{C}_2 \\ \hline \mathbf{C}_1 & \mathbf{C}_3 \end{array}\right] \tag{2-43}$$

(5) check to see that $\mathbf{C}_1$ represents the column sums of **C**; (6) verify the fact that $\mathbf{C}_2$ represents the row sums of **C**; (7) check to see that $\mathbf{C}_3$ represents the sum of the elements in both $\mathbf{C}_1$ and $\mathbf{C}_2$; (8) then, if all checks have been satisfied, the matrix **C** may be assumed to be calculated correctly and equal to the product **AB**.

Problems

2.2-1. Determine the matrix **C** given by the following expression:

$$\mathbf{C} = 3\mathbf{A} - 2\mathbf{B}$$

if the matrices **A** and **B** are

$$\mathbf{A} = \begin{bmatrix} 2 & -1 \\ 0 & 3 \\ -4 & 1 \end{bmatrix} \qquad \mathbf{B} = \begin{bmatrix} 4 & -3 \\ 1 & 2 \\ -2 & 5 \end{bmatrix}$$

2.2-2. Obtain the matrix **C** in order that the following equation is satisfied:

$$-5\mathbf{A} + 2\mathbf{C} = \mathbf{B}$$

if the matrices **A** and **B** are

$$\mathbf{A} = \begin{bmatrix} 7 & -2 \\ 4 & 5 \end{bmatrix} \qquad \mathbf{B} = \begin{bmatrix} -3 & 4 \\ 2 & 5 \end{bmatrix}$$

2.4-1. Given the matrices **A** and **B** as shown below, find the product **AB**.

$$\mathbf{A} = \begin{bmatrix} 4 & 0 & -2 & 1 \\ 3 & -2 & 4 & 3 \end{bmatrix} \qquad \mathbf{B} = \begin{bmatrix} 3 \\ -2 \\ 1 \\ 4 \end{bmatrix}$$

2.4-2. Premultiply the matrix **B** given in the preceding problem by the following matrix:

$$\mathbf{A} = \begin{bmatrix} -2 & 1 & 0 & 1 \\ 3 & -1 & 2 & 4 \\ 0 & -2 & 0 & -3 \end{bmatrix}$$

2.4–3. Find the product **BC** of the matrix **B** given in Problem 2.4–1 and the matrix

$$\mathbf{C} = [2 \quad -2 \quad 3]$$

2.4–4. Given the matrices **D** and **E** as shown below, find the products **DE** and **ED**.

$$\mathbf{D} = \begin{bmatrix} -1 & 2 & 2 & 6 \\ 7 & -3 & -4 & 0 \end{bmatrix} \qquad \mathbf{E} = \begin{bmatrix} 6 & 3 \\ -1 & 0 \\ 0 & -4 \\ 2 & 1 \end{bmatrix}$$

2.4–5. Under what conditions will the multiplication of two matrices produce a product matrix that is (a) of order 1 × 1, (b) a column matrix, (c) a row matrix, and (d) a square matrix?

2.4–6. Postmultiply the matrix **A** of Problem 2.4–1 by the matrix **E** of Problem 2.4–4.

2.4–7. Premultiply the matrix **A** of Problem 2.4–1 by the matrix **E** of Problem 2.4–4.

2.4–8. Calculate the products **FG** and **GF** for the following square matrices:

$$\mathbf{F} = \begin{bmatrix} 2 & -1 \\ -3 & 4 \end{bmatrix} \qquad \mathbf{G} = \begin{bmatrix} 1 & 4 \\ -1 & 1 \end{bmatrix}$$

2.4–9. Find the products **HJ** and **JH** for the following square matrices:

$$\mathbf{H} = \begin{bmatrix} 1 & 3 & 0 \\ 2 & -1 & 0 \\ 3 & 2 & 0 \end{bmatrix} \qquad \mathbf{J} = \begin{bmatrix} 0 & 0 & 0 \\ 0 & 0 & 0 \\ 2 & 5 & 7 \end{bmatrix}$$

2.4–10. Referring to the matrices **A**, **B**, and **C** of Problems 2.4–1 and 2.4–3, show that $(\mathbf{AB})\mathbf{C} = \mathbf{A}(\mathbf{BC})$.

2.4–11. Referring to the matrices **A**, **B**, and **D** of Problems 2.4–1 and 2.4–4, show that $(\mathbf{A} + \mathbf{D})\mathbf{B} = \mathbf{AB} + \mathbf{DB}$.

2.5–1. Using the matrices $\mathbf{A}_1$ and $\mathbf{A}_2$ given below, demonstrate that the transpose of the sum of two matrices is the sum of the transposes (see Equation 2–18a) and that the transpose of the product of two matrices is the product of the transposes in reverse order (see Equation 2–19a).

$$\mathbf{A}_1 = \begin{bmatrix} 1 & 0 & 2 \\ -3 & 2 & 1 \\ -4 & 1 & 3 \end{bmatrix} \qquad \mathbf{A}_2 = \begin{bmatrix} 5 & -2 & 0 \\ 1 & -1 & 2 \\ 3 & 0 & 4 \end{bmatrix}$$

2.5–2. Using the matrices $\mathbf{B}_1$, $\mathbf{B}_2$, and $\mathbf{B}_3$ given below, show that the following relation (see Equation 2–19b) is satisfied:

$$(\mathbf{B}_1\mathbf{B}_2\mathbf{B}_3)^{\mathsf{T}} = \mathbf{B}_3^{\mathsf{T}}\mathbf{B}_2^{\mathsf{T}}\mathbf{B}_1^{\mathsf{T}}$$

$$\mathbf{B}_1 = \begin{bmatrix} -4 & 1 \\ 2 & 3 \end{bmatrix} \qquad \mathbf{B}_2 = \begin{bmatrix} 2 & 1 \\ 0 & -3 \end{bmatrix} \qquad \mathbf{B}_3 = \begin{bmatrix} 3 & -2 \\ -1 & 4 \end{bmatrix}$$

2.5–3. Find the products $\mathbf{D}_1\mathbf{D}_2$ and $\mathbf{D}_2\mathbf{D}_1$ for the following diagonal matrices:

$$\mathbf{D}_1 = \begin{bmatrix} 2 & 0 & 0 \\ 0 & 3 & 0 \\ 0 & 0 & -1 \end{bmatrix} \qquad \mathbf{D}_2 = \begin{bmatrix} -4 & 0 & 0 \\ 0 & 3 & 0 \\ 0 & 0 & 1 \end{bmatrix}$$

2.5–4. What is the result of premiltiplying the matrix $\mathbf{A}_1$ of Problem 2.5–1 by the diagonal matrix $\mathbf{D}_1$ of Problem 2.5–3? What is the result of postmultiplying?

2.5–5. Show that any square matrix $\mathbf{A}$ can be expressed as the sum of a symmetric and a skew-symmetric matrix. (Hint: Begin by showing that $\mathbf{A} + \mathbf{A}^\mathrm{T}$ is symmetric and $\mathbf{A} - \mathbf{A}^\mathrm{T}$ is skew-symmetric.)

2.5–6. Show that when any rectangular or square matrix $\mathbf{A}$ is pre- or post-multiplied by its transpose $\mathbf{A}^\mathrm{T}$ the product is a symmetric matrix.

2.5–7. Find the square and cube of the matrix $\mathbf{D}_1$ given in Problem 2.5–3.

2.5–8. Find the square and cube of the matrix $\mathbf{A}_1$ given in Problem 2.5–1.

2.5–9. Show that the following two matrices and their negatives are square roots of the same matrix:

$$\begin{bmatrix} 1 & 2 \\ 2 & -2 \end{bmatrix} \qquad \frac{1}{5}\begin{bmatrix} 11 & -2 \\ -2 & 14 \end{bmatrix}$$

2.5–10. Show that the following four matrices and their negatives are square roots of the same matrix:

$$\frac{1}{\sqrt{11}}\begin{bmatrix} -33 & 11 & 0 \\ 12 & -26 & 6 \\ 8 & 23 & -29 \end{bmatrix} \qquad \frac{1}{\sqrt{11}}\begin{bmatrix} -29 & 17 & 2 \\ 20 & -14 & 10 \\ 20 & 41 & -23 \end{bmatrix}$$

$$\frac{1}{\sqrt{11}}\begin{bmatrix} 3 & -23 & 18 \\ -12 & 26 & -6 \\ 52 & 1 & -7 \end{bmatrix} \qquad \frac{1}{\sqrt{11}}\begin{bmatrix} 1 & 29 & -16 \\ 20 & -14 & 10 \\ -40 & 17 & 13 \end{bmatrix}$$

In solving the problems for Section 2.6, construct either the elementary matrices or the transformation matrices that are required; show the order in which the matrices are to be multiplied; and perform the matrix multiplications.

2.6–1. Given the matrix $\mathbf{A}$ (see below), multiply the second row by three and change the signs of the second and third columns.

$$\mathbf{A} = \begin{bmatrix} 2 & -1 & -3 \\ -4 & 2 & 1 \end{bmatrix}$$

2.6–2. Given the matrix $\mathbf{B}$ (see below), interchange the second and fourth rows and the first and second columns.

$$\mathbf{B} = \begin{bmatrix} -2 & 5 & 2 \\ 1 & -6 & 0 \\ 0 & 1 & -3 \\ 3 & 4 & -1 \end{bmatrix}$$

2.6–3. Referring to matrix **A** of Problem 2.6–1, subtract the first row from the second row, and add twice the third column to the first column.

2.6–4. Referring to matrix **B** of Problem 2.6–2, add the second row to the third row, subtract the fourth row from the third row, and subtract three times the second column from the third column.

2.6–5. Given the matrix **C** (see below), subtract twice the first row from the second row, multiply the second row by two, multiply the third row by five, interchange the first and second columns, and multiply the first column by two.

$$\mathbf{C} = \begin{bmatrix} 1 & 4 & 0 \\ -2 & 2 & -1 \\ 6 & 4 & 3 \end{bmatrix}$$

In solving the problems for Section 2.7, obtain the results by multiplication of submatrices.

2.7–1. Calculate the product **AB** of the matrices shown below, using the partitioning indicated.

$$\mathbf{A} = \left[\begin{array}{cc|c} 1 & 0 & 2 \\ -3 & 2 & 1 \end{array}\right] \qquad \mathbf{B} = \left[\begin{array}{cc} 2 & 1 \\ 3 & 4 \\ \hline 0 & -2 \end{array}\right]$$

2.7–2. Given the matrices **C** and **D** (see below), calculate the product **CD** using the partitioning shown.

$$\mathbf{C} = \left[\begin{array}{cc|cc} 2 & 1 & 0 & 3 \\ -1 & 0 & 1 & 1 \\ \hline -1 & 2 & -3 & 0 \\ 0 & 4 & 1 & 2 \end{array}\right] \qquad \mathbf{D} = \left[\begin{array}{cc|cc} 5 & 0 & 1 & -1 \\ 0 & 1 & -2 & 0 \\ \hline -5 & 1 & 2 & -1 \\ 0 & 2 & 1 & 1 \end{array}\right]$$

2.7–3. Given the matrices **E** and **F** (see below), calculate the product **EF** by partitioning **E** into four submatrices and **F** into two submatrices.

$$\mathbf{E} = \begin{bmatrix} 2 & 1 & 0 & -2 & 0 & 3 \\ -1 & 0 & 4 & 0 & 1 & 0 \\ 0 & 3 & 0 & -4 & 0 & 1 \end{bmatrix} \qquad \mathbf{F} = \begin{bmatrix} 4 & 0 & -1 & 0 \\ 0 & 0 & 2 & 1 \\ 3 & -1 & 0 & 0 \\ 0 & 1 & 3 & 0 \\ 2 & 0 & 0 & 1 \\ 0 & 2 & 0 & 2 \end{bmatrix}$$

2.7–4. Solve the preceding problem by partitioning both **E** and **F** into four submatrices.

2.7–5. Calculate the square of the following matrix using the partitioning shown.

$$\mathbf{G} = \left[\begin{array}{ccc|cc} 1 & 0 & 1 & 0 & 1 \\ 2 & 1 & 0 & 2 & 1 \\ 0 & 0 & 1 & 0 & 1 \\ \hline 1 & 1 & 0 & 1 & 1 \\ 0 & 2 & 0 & 0 & 1 \end{array}\right]$$

In solving the problems for Section 2.8, the checking procedure is to be used in conjunction with the matrix multiplication.

2.8–1. Calculate the product **AB** for the matrices shown.

$$\mathbf{A} = \begin{bmatrix} 2 & 3 \\ 1 & 0 \end{bmatrix} \qquad \mathbf{B} = \begin{bmatrix} 1 & 4 \\ 0 & 1 \end{bmatrix}$$

2.8–2. Given the matrices **C** and **D** (see below), obtain the product **CD**.

$$\mathbf{C} = \begin{bmatrix} 5 & -5 & 10 \\ 10 & 0 & 20 \end{bmatrix} \qquad \mathbf{D} = \begin{bmatrix} 1 & 2 \\ -2 & 0 \\ 1 & 1 \end{bmatrix}$$

2.8–3. Calculate the product **AB** for the matrices given in Problem 2.4–1.

2.8–4. Calculate the product **DE** for the matrices given in Problem 2.4–4.

2.8–5. Calculate the square and cube of the matrix **A** in Problem 2.8–1.

3

...Determinants...

3.1. Definitions. Determinants serve a useful role in the process of inversion of matrices, in the solution of simultaneous equations, and in obtaining the characteristic equation for an eigenvalue problem (see Chapters 4, 5, and 6, respectively). Therefore, some of the properties of determinants, as well as methods for evaluating them, are given in this chapter.

Determinants arise quite naturally in the solution of simultaneous equations. For instance, consider two simultaneous equations with two unknowns X_1 and X_2:

$$A_{11}X_1 + A_{12}X_2 = B_1$$

$$A_{21}X_1 + A_{22}X_2 = B_2$$

In matrix form the equations are

$$\begin{bmatrix} A_{11} & A_{12} \\ A_{21} & A_{22} \end{bmatrix} \begin{bmatrix} X_1 \\ X_2 \end{bmatrix} = \begin{bmatrix} B_1 \\ B_2 \end{bmatrix} \tag{a}$$

The solution of these equations for X_1 and X_2 is

$$X_1 = \frac{B_1A_{22} - A_{12}B_2}{A_{11}A_{22} - A_{12}A_{21}} \tag{b}$$

and

$$X_2 = \frac{A_{11}B_2 - B_1A_{21}}{A_{11}A_{22} - A_{12}A_{21}} \tag{c}$$

as can be verified readily by substituting into the original equations. Of particular interest is the fact that the denominator terms in Equations (b) and (c) are the same. Each denominator consists of the product of the elements on the principal diagonal of the 2×2 matrix of coefficients **A** minus the product of the elements on the secondary diagonal. This concept is formalized by enclosing the elements of the matrix **A** between vertical lines and referring to them in this form as the *determinant* of **A**, which has the expansion formula shown:

$$\begin{vmatrix} A_{11} & A_{12} \\ A_{21} & A_{22} \end{vmatrix} = A_{11}A_{22} - A_{12}A_{21} \tag{3-1}$$

Using this definition of a 2 X 2 determinant, it is possible to express not only the denominators, but also the numerators, of Equations (b) and (c) in determinantal form. This form of solution can be extended to include any number of simultaneous equations and is described later in Section 5.3 as Cramer's rule.

From the preceding discussion of two simultaneous equations, it can be seen that the concept of a determinant originates with the solution of simultaneous equations. However, in more general terms, a determinant can be defined as a square array of numbers that is evaluated (or expanded) according to established mathematical rules. These rules, which are described in the remainder of this chapter, are based upon the relationship of determinants to the solution of simultaneous equations, as illustrated in Equation (3-1) for a 2 X 2 determinant.

The *order* of a determinant refers to the number of rows or columns in the determinant; thus, the determinant in Equation (3-1) is of order two. It is possible to form from any square matrix a determinant of the same order as the matrix. If the matrix is denoted as **A**, then the determinant formed from its elements is called the *determinant of the matrix* and is denoted either det(**A**) or |**A**|. This notation is illustrated by the following example, in which it is assumed that **A** is a square matrix of order three:

$$\mathbf{A} = \begin{bmatrix} A_{11} & A_{12} & A_{13} \\ A_{21} & A_{22} & A_{23} \\ A_{31} & A_{32} & A_{33} \end{bmatrix}$$

Then the determinant of **A** is written as

$$\det(\mathbf{A}) = |\mathbf{A}| = \begin{vmatrix} A_{11} & A_{12} & A_{13} \\ A_{21} & A_{22} & A_{23} \\ A_{31} & A_{32} & A_{33} \end{vmatrix} \tag{3-2}$$

The quantities $A_{11}, A_{12}, \ldots$, etc., are called the *elements* of the determinant. If the elements are numbers, the expansion of the determinant will produce a single numerical value. On the other hand, if the elements of the determinant are other mathematical quantities, then the expansion of the determinant will produce a polynomial containing those quantities.

The rules for expanding determinants derive from the form of the denominator terms in equations such as Equations (b) and (c) for the solution of two simultaneous equations. Thus, the rule for expansion of a determinant of order two is given by Equation (3-1). This expansion is represented schematically by the arrows shown in Figure 3-1a, where the sign to be used with the product is indicated at the tail of the arrow. It may also be shown that a third-order determinant is evaluated as follows:

$$\begin{vmatrix} A_{11} & A_{12} & A_{13} \\ A_{21} & A_{22} & A_{23} \\ A_{31} & A_{32} & A_{33} \end{vmatrix} = A_{11}A_{22}A_{33} + A_{12}A_{23}A_{31} + A_{13}A_{21}A_{32} - A_{13}A_{22}A_{31} - A_{11}A_{23}A_{32} - A_{12}A_{21}A_{33} \qquad (3\text{-}3)$$

Although the expression in Equation (3-3) seems rather long, it is not necessary to memorize it since it can be obtained readily from the diagram shown in Figure 3-1b. In the diagram, the first two columns of the determinant are repeated at the right, and then triple products are formed along the arrows; those formed by arrows from upper left to lower right are taken positive; those formed by arrows from upper right to lower left are negative. The methods of evaluation shown in Figures 3-1a and b are suitable only for determinants of second and third order, respectively, and cannot be used for higher order determinants. Such determinants have expansions which become increasingly complicated with an increase in their order, and they must be expanded by methods such as those described in the remainder of this chapter.

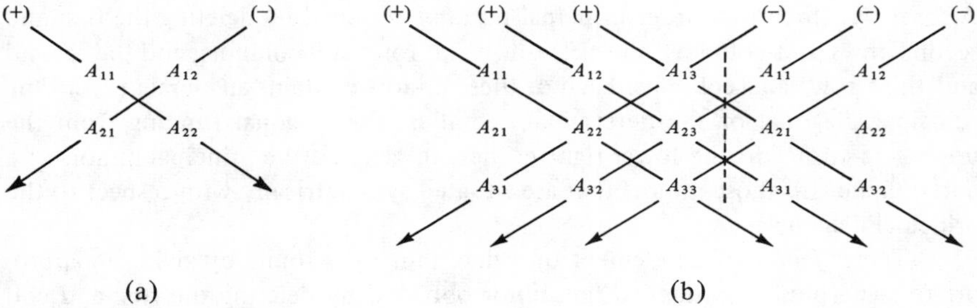

Figure 3-1. Expansion rules for 2 X 2 and 3 X 3 determinants

3.2. Minors and Cofactors. A *minor* of a determinant is another determinant formed by removing an equal number of rows and columns from the original determinant. The *order* of the minor refers to the number of rows or columns in the minor determinant itself. As an illustration, a determinant formed from a fourth-order determinant by removing one row and one column is called a minor of third order; if two rows and columns are removed, there will remain a minor of second order; and so on. In the case of the third-order determinant in Equation (3-2), it is possible to obtain a second-order minor by deleting the first row of the determinant and any one of the columns. Thus, by deleting the first row and columns one, two, and three, respectively, the following three second-order minors are obtained:

$$\begin{vmatrix} A_{22} & A_{23} \\ A_{32} & A_{33} \end{vmatrix} \quad \begin{vmatrix} A_{21} & A_{23} \\ A_{31} & A_{33} \end{vmatrix} \quad \begin{vmatrix} A_{21} & A_{22} \\ A_{31} & A_{32} \end{vmatrix}$$

In a similar manner, three second-order minors are obtained by deletion of row two and each of the three columns, and three more minors may be found by deletion of row three and each of the three columns. Thus, a determinant of order three is seen to have nine second-order minors. The first-order minors of a third-order determinant are formed by deleting two rows and two columns. Each such minor is a determinant containing only one element, and it is apparent that there are nine first-order minors.

A *principal minor* of a determinant is formed when the rows and columns that are removed are in the same relative positions. Thus, the third-order determinant of Equation (3-2) has the following three second-order principal minors, obtained by deleting the first row and first column, the second row and second column, and the third row and third column, respectively:

$$\begin{vmatrix} A_{22} & A_{23} \\ A_{32} & A_{33} \end{vmatrix} \quad \begin{vmatrix} A_{11} & A_{13} \\ A_{31} & A_{33} \end{vmatrix} \quad \begin{vmatrix} A_{11} & A_{12} \\ A_{21} & A_{22} \end{vmatrix}$$

Also, it has three first-order principal minors, obtained by deleting the first and second rows and columns, the first and third rows and columns, and the second and third rows and columns. Each of these minors contains an element from the principal diagonal of the determinant, that is, the diagonal running from the upper-left corner to the lower-right corner. In general, the principal minors of a determinant are those minors that are situated symmetrically with respect to the principal diagonal.

The *cofactor* of an element of a determinant is found by giving an appropriate sign (plus or minus) to the minor obtained by deleting the row and column containing that element. The sign to be used depends upon the position of the element in the determinant. If the sum of the number of the row and the number of the column containing the element is an even number, the sign attached to the minor is positive; if this sum is an odd number, the sign is negative. For example, if the element is A_{22}, the sign is taken positive; if the element is A_{32}, the sign is negative. In general, the cofactor of element A_{ij} is equal to $(-1)^{i+j}$ times the minor corresponding to A_{ij}. This rule is easy to remember because it produces a checkerboard pattern of plus and minus signs (see Figure 3-2). Sometimes cofactors are called *signed minors*. The order of the minor used in obtaining a cofactor is always one less than the order of the original determinant. The cofactor of an element A_{ij} of a determinant will be denoted as A_{ij}^c.

$$\begin{vmatrix} + & - & + & - & \cdots \\ - & + & - & + & \cdots \\ + & - & + & - & \cdots \\ - & + & - & + & \cdots \\ \cdots & & & & \cdots \end{vmatrix}$$

Figure 3-2. Checkerboard pattern of signs for obtaining cofactors

Again using the determinant in Equation (3-2) as an example, it is seen that the cofactor of element A_{11} is

$$A_{11}^{c} = \begin{vmatrix} A_{22} & A_{23} \\ A_{32} & A_{33} \end{vmatrix} = A_{22}A_{33} - A_{23}A_{32} \tag{a}$$

Also, the cofactor of element A_{12} is

$$A_{12}^{c} = -\begin{vmatrix} A_{21} & A_{23} \\ A_{31} & A_{33} \end{vmatrix} = -(A_{21}A_{33} - A_{23}A_{31}) \tag{b}$$

and the cofactor of element A_{13} is

$$A_{13}^{c} = \begin{vmatrix} A_{21} & A_{22} \\ A_{31} & A_{32} \end{vmatrix} = A_{21}A_{32} - A_{22}A_{31} \tag{c}$$

and similarly for the other elements.

It is possible to evaluate any determinant through the use of cofactors of a row or column. In this method, any row (or column) of the determinant may be selected, and then each element in the row (or column) is multiplied by its cofactor. The sum of these products gives the value of the determinant. In order to demonstrate this method, which is called *expansion by cofactors,** consider the evaluation of the third-order determinant shown in Equation (3-3). The expansion in terms of the cofactors of elements of the first row (see Equations a, b, and c) may be written as follows:

$$\begin{aligned} |\mathbf{A}| &= A_{11}A_{11}^{c} + A_{12}A_{12}^{c} + A_{13}A_{13}^{c} \\ &= A_{11}(A_{22}A_{33} - A_{23}A_{32}) - A_{12}(A_{21}A_{33} - A_{23}A_{31}) \\ &\quad + A_{13}(A_{21}A_{32} - A_{22}A_{31}) \end{aligned}$$

which agrees with Equation (3-3). The same expression for the determinant is obtained no matter which row or column is used in the expansion. The method

*The method of expansion by cofactors is a specialized form of a more general method for evaluation known as the *Laplacian expansion*. (Pierre Simon Laplace, 1749–1827, was a French mathematician.)

can be used for determinants of any order, but if the cofactors themselves contain large determinants, the process must be repeated many times until the final value of the original determinant is obtained. Thus, the method is cumbersome if used for very large determinants, and other methods of expansion are preferable.

As a numerical example illustrating the expansion of a determinant by cofactors, consider the fourth-order determinant shown below:

$$D = \begin{vmatrix} 2 & -3 & 1 & 2 \\ 1 & 3 & -4 & 0 \\ 0 & 1 & -1 & 2 \\ 5 & -3 & 0 & 1 \end{vmatrix}$$

in which the letter D is used as a convenient symbol for an arbitrarily selected determinant. When the determinant D is expanded by elements of the first row, each element of the first row is multiplied by its cofactor and then the terms are added, as follows:

$$D = 2\begin{vmatrix} 3 & -4 & 0 \\ 1 & -1 & 2 \\ -3 & 0 & 1 \end{vmatrix} + 3\begin{vmatrix} 1 & -4 & 0 \\ 0 & -1 & 2 \\ 5 & 0 & 1 \end{vmatrix}$$

$$+\begin{vmatrix} 1 & 3 & 0 \\ 0 & 1 & 2 \\ 5 & -3 & 1 \end{vmatrix} - 2\begin{vmatrix} 1 & 3 & -4 \\ 0 & 1 & -1 \\ 5 & -3 & 0 \end{vmatrix}$$

Each of the third-order determinants (or minors) in this expression can be expanded, either by cofactors or by the rule illustrated in Figure 3-1b. The final results are as follows:

$$D = 2(25) + 3(-41) + 37 - 2(2) = -40$$

The same value for the determinant will be obtained if it is expanded by the cofactors of the elements of any other row or any column. If the selected row or column has some zero elements, the number of minor determinants to be evaluated will be reduced correspondingly. This idea forms the basis for the method of expansion known as *pivotal condensation*, which is described in Section 3.4.

3.3. Properties of Determinants. Determinants have many important properties that are useful when performing calculations. A few of these properties have been mentioned in previous sections; many others are explained in this section. For instance, an important property is the following. If any two rows or columns of a determinant are interchanged, the value of the determinant will change sign. This property can be established for the case when the two rows (or columns) are adjacent simply by expanding the determinant using cofactors of the elements of one of the rows or columns that are interchanged. Since every cofactor merely changes sign (see Figure 3-2), the value of the determinant will change sign also. To illustrate this fact for a third-order determinant, consider the expansion by elements of the first row:

$$\begin{vmatrix} A_{11} & A_{12} & A_{13} \\ A_{21} & A_{22} & A_{23} \\ A_{31} & A_{32} & A_{33} \end{vmatrix} = A_{11}\begin{vmatrix} A_{22} & A_{23} \\ A_{32} & A_{33} \end{vmatrix} - A_{12}\begin{vmatrix} A_{21} & A_{23} \\ A_{31} & A_{33} \end{vmatrix} + A_{13}\begin{vmatrix} A_{21} & A_{22} \\ A_{31} & A_{32} \end{vmatrix} \tag{3-4a}$$

If the first and second rows of the determinant are interchanged, the expansion by elements of the second row becomes

$$\begin{vmatrix} A_{21} & A_{22} & A_{23} \\ A_{11} & A_{12} & A_{13} \\ A_{31} & A_{32} & A_{33} \end{vmatrix} = -A_{11}\begin{vmatrix} A_{22} & A_{23} \\ A_{32} & A_{33} \end{vmatrix} + A_{12}\begin{vmatrix} A_{21} & A_{23} \\ A_{31} & A_{33} \end{vmatrix} - A_{13}\begin{vmatrix} A_{21} & A_{22} \\ A_{31} & A_{32} \end{vmatrix} \tag{3-4b}$$

Inspection of Equations (3-4a) and (3-4b) shows that the latter is the same as the former, except for a change in sign. If the two rows (or columns) being interchanged are not adjacent, then it is always possible to arrive at the desired interchange by successively interchanging adjacent rows (or columns). Each such adjacent interchange means a change in sign of the determinant, and since the number of interchanges is always an odd number, the final result is a change in sign of the determinant. For instance, suppose rows 2 and 4 of a determinant are to be interchanged. The successive adjacent interchanges are as follows: interchange rows 2 and 3, interchange (new) rows 3 and 4, and interchange (new) rows 2 and 3. These interchanges accomplish the desired result of interchanging rows 2 and 4 of the original determinant. Three adjacent interchanges were required, each changing the sign of the determinant, hence the final result is that the determinant has changed sign.

Next, consider the result of evaluating the determinant of a transposed square matrix. For a matrix of order two, the determinants of the matrix and its transpose are

$$|\mathbf{B}| = \begin{vmatrix} B_{11} & B_{12} \\ B_{21} & B_{22} \end{vmatrix} = B_{11}B_{22} - B_{12}B_{21}$$

$$|\mathbf{B}^{\mathbf{T}}| = \begin{vmatrix} B_{11} & B_{21} \\ B_{12} & B_{22} \end{vmatrix} = B_{11}B_{22} - B_{21}B_{12}$$

The above expansions for the second-order determinants show that they have the same value. For a determinant of order three, an expansion by the first row was shown previously in Equation (3–4a). If the corresponding matrix is transposed, an expansion of the determinant of the matrix by elements of the first column gives

$$\begin{vmatrix} A_{11} & A_{21} & A_{31} \\ A_{12} & A_{22} & A_{32} \\ A_{13} & A_{23} & A_{33} \end{vmatrix} = A_{11} \begin{vmatrix} A_{22} & A_{32} \\ A_{23} & A_{33} \end{vmatrix} - A_{12} \begin{vmatrix} A_{21} & A_{31} \\ A_{23} & A_{33} \end{vmatrix} + A_{13} \begin{vmatrix} A_{21} & A_{31} \\ A_{22} & A_{32} \end{vmatrix}$$

This equation is the same as Equation (3–4a) except that each of the second-order determinants has been transposed. But it has already been observed that such second-order transposed determinants are equal; hence the conclusion is reached that the same equality holds for a third-order determinant. By taking a row expansion of $|\mathbf{A}|$ and a column expansion of $|\mathbf{A}^{\mathrm{T}}|$, where $\mathbf{A}$ is a square matrix of any order, it can be shown that the determinants are equal in all cases. Thus the following general rule is obtained:

$$|\mathbf{A}| = |\mathbf{A}^{\mathrm{T}}| \tag{3-5}$$

This equation states that the value of a determinant is not changed by transposition.

When all elements of one row or column of a determinant are multiplied by a scalar number λ, the value of the determinant is multiplied by λ also. This property is evident from an expansion by cofactors using elements of the row or column that has been multiplied by λ. An extension of this conclusion shows that if the elements of any two rows or any two columns are multiplied by λ, the determinant is multiplied by λ^2; if three rows or columns are multiplied by λ, the determinant is multiplied by λ^3, and so forth. If the determinant is of order n and if all elements of the determinant are multiplied by λ, the value of the determinant will be multiplied by λ^n. Since multiplication of a matrix by a scalar means that every element in the matrix is multiplied by the scalar, it follows that

$$|\lambda\mathbf{A}| = \lambda^n |\mathbf{A}| \tag{3-6}$$

where $\mathbf{A}$ is a square matrix of order n.

If two square matrices $\mathbf{A}$ and $\mathbf{B}$ of the same order are multiplied, the determinant of their products is equal to the product of their determinants; that is,

$$|\mathbf{AB}| = |\mathbf{BA}| = |\mathbf{A}|\,|\mathbf{B}| \tag{3-7}$$

Of course, the matrix products **AB** and **BA** are not necessarily the same, as explained previously in Section 2.3; only their determinants are equal. The proof of this theorem is complicated, but its validity can easily be demonstrated by examples (see Problems 3.3-3 and 3.3-4). An obvious generalization of Equation (3-7) is the following:

$$|\mathbf{ABC}\ldots| = |\mathbf{A}|\,|\mathbf{B}|\,|\mathbf{C}|\ldots \tag{3-8}$$

that is, the determinant of the product of any number of square matrices of the same order is equal to the product of their determinants.

A special case of the preceding multiplication rules occurs when one of the matrices has a determinant equal to zero; in that case, the matrix product will also have a zero determinant. A square matrix having its determinant equal to zero is said to be *singular*; singular matrices are discussed further in Section 4.4. If the product of two or more square matrices yields the null matrix, as shown:

$$\mathbf{ABC}\ldots = \mathbf{0} \tag{3-9a}$$

then at least one of the matrices must be singular. If there are only two matrices in the product, that is, if

$$\mathbf{AB} = \mathbf{0} \tag{3-9b}$$

where **A** and **B** are square matrices and neither **A** nor **B** is a null matrix, then both **A** and **B** are singular. This rule is demonstrated in Example 5 of Section 2.4.

Another interesting rule pertains to the determinant of the product of two rectangular matrices. Suppose that **A** is a rectangular matrix of order $m \times n$ and **B** is a rectangular matrix of order $n \times m$. Then, if $m > n$, the determinant of the product **AB**, which is a square matrix of order m, must be equal to zero, and hence the product is a singular matrix:

$$|\mathbf{AB}| = 0 \quad \text{for } m > n \tag{3-10}$$

To establish this theorem in general, it is only necessary to add $m - n$ columns of zeroes to the matrix **A** and $m - n$ rows of zeros to **B** (thus making both matrices of order $m \times m$), multiply them, and use Equation (3-7).

As a precaution against inferring an incorrect result, it should be observed that the determinant of the sum of two square matrices is not generally equal to the sum of the determinants of the matrices (see Problem 3.3-7).

The determinant of a diagonal matrix $\mathbf{D}_n$ is clearly equal to the product of the diagonal elements:

$$\mathbf{D}_n = \begin{bmatrix} D_{11} & 0 & \ldots & 0 \\ 0 & D_{22} & \ldots & 0 \\ \ldots & \ldots & \ldots & \ldots \\ 0 & 0 & \ldots & D_{nn} \end{bmatrix} \tag{3-11}$$

$$|\mathbf{D}_n| = D_{11}D_{22}\ldots D_{nn} \tag{3-12}$$

Furthermore, the same rule also holds for a triangular matrix (see Problem 3.2-9).

If a square matrix has nonzero elements only on the secondary diagonal, its determinant is equal to the product of the secondary diagonal elements if the order of the matrix is $n = 1, 4, 5, 8, 9, 12, 13, \ldots$; otherwise, it is equal to the negative of that product. This rule can be easily established by direct evaluation of the determinants.

In the case of a skew-symmetric matrix $\mathbf{A}$ of order n, it is not difficult to show that the determinant of the matrix must equal zero if n is odd. The definition of a skew-symmetric matrix gives $\mathbf{A} = -\mathbf{A}^{\mathsf{T}}$, from which

$$|\mathbf{A}| = |-\mathbf{A}^{\mathsf{T}}| \tag{a}$$

Equations (3-6) and (3-5) now yield

$$|-\mathbf{A}^{\mathsf{T}}| = (-1)^n|\mathbf{A}^{\mathsf{T}}| = (-1)^n|\mathbf{A}| \tag{b}$$

Combining Equations (a) and (b) shows that

$$|\mathbf{A}| = (-1)^n|\mathbf{A}|$$

If n is even, this equation becomes an identity and no conclusions can be drawn; however, if n is odd, the determinant of $\mathbf{A}$ is equal to its negative, which means that the determinant must equal zero.

If a square matrix $\mathbf{D}$ can be partitioned into submatrices in the following manner:

$$\mathbf{D} = \begin{bmatrix} \mathbf{A} & \mathbf{0} \\ \mathbf{B} & \mathbf{C} \end{bmatrix} \tag{3-13}$$

in which $\mathbf{A}$ and $\mathbf{C}$ are square matrices (not necessarily of the same order), then the determinant of $\mathbf{D}$ is equal to the product of the determinant of $\mathbf{A}$ and the determinant of $\mathbf{C}$; thus,

$$|\mathbf{D}| = |\mathbf{A}|\,|\mathbf{C}| \tag{3-14}$$

This rule is demonstrated in Problem 3.3-8.

There are several conditions under which the value of a determinant will be zero. The most evident one is when all elements of a row or column are zero. In such a case, an expansion by cofactors using the row or column having zero elements will give a value of zero for the determinant. The same result will occur if all the cofactors of the elements of any row or column are zero. Further, if two rows or columns of a determinant are identical, the determinant will be equal to zero. This conclusion can be seen by interchanging the two identical rows or columns, which must change the sign of the determinant. However, because the determinant itself has not changed, there can be no change in its value; such a result is possible only if the determinant has the value zero. The same result is found if two rows or columns are proportional; that is, if all elements of one row or column are scalar multiples of the corresponding elements of another row or column. By factoring from the row or column the scalar multiple and placing it outside of the determinant, a determinant will remain that has two identical rows or columns, and hence it is equal to zero.

The condition just described, in which one row (or column) of a determinant is a scalar multiple of another row (or column), is a special case of a more general situation in which the rows (or columns) of a determinant are linearly dependent. Linear dependence is discussed later in Section 5.6, but in brief it means that one row (or column) of the determinant can be expressed as a linear combination of the other rows (or columns). It can be shown that the value of a determinant is zero if its rows (or columns) are linearly dependent.

In the expansion of a determinant by cofactors (see Section 3.2), the elements of one row (or column) are multiplied by their cofactors and then added; the result is the value of the determinant. This property has been used often in the preceding discussions. A related property that will be used in the next chapter in connection with finding the inverse of a matrix is the following. If the elements of one row (or column) of a determinant are multiplied by cofactors of the corresponding elements of a different row (or column) and then added, the result is always zero. To illustrate this statement, consider again the determinant of order three:

$$|\mathbf{A}| = \begin{vmatrix} A_{11} & A_{12} & A_{13} \\ A_{21} & A_{22} & A_{23} \\ A_{31} & A_{32} & A_{33} \end{vmatrix} \tag{3-15}$$

Now suppose, as an example, that the elements of the second row of this determinant are multiplied by cofactors of the corresponding elements of the third row and then added. The result is

$$A_{21}\begin{vmatrix} A_{12} & A_{13} \\ A_{22} & A_{23} \end{vmatrix} - A_{22}\begin{vmatrix} A_{11} & A_{13} \\ A_{21} & A_{23} \end{vmatrix} + A_{23}\begin{vmatrix} A_{11} & A_{12} \\ A_{21} & A_{22} \end{vmatrix}$$

or

$$A_{21}(A_{12}A_{23} - A_{13}A_{22}) - A_{22}(A_{11}A_{23} - A_{13}A_{21}) + A_{23}(A_{11}A_{22} - A_{12}A_{21})$$

Upon expanding this last expression it is found that all of the terms cancel, and hence the value of the expression is zero. Thus, it has been shown that when elements of the second row of $|\mathbf{A}|$ are multiplied by cofactors of the third row, the result is zero. In order to establish this property in general, consider any nth order determinant:

$$D_1 = \begin{vmatrix} A_{11} & A_{12} & \dots & A_{1n} \\ \dots & \dots & \dots & \dots \\ A_{i1} & A_{i2} & \dots & A_{in} \\ \dots & \dots & \dots & \dots \\ A_{k1} & A_{k2} & \dots & A_{kn} \\ \dots & \dots & \dots & \dots \\ A_{n1} & A_{n2} & \dots & A_{nn} \end{vmatrix} \tag{3-16}$$

If elements of the ith row of this determinant are multiplied by cofactors of the corresponding elements in the kth row, the result is exactly the same as the expansion of a determinant D_2 in which the kth row is the same as the ith row:

$$D_2 = \begin{vmatrix} A_{11} & A_{12} & \dots & A_{1n} \\ \dots & \dots & \dots & \dots \\ A_{i1} & A_{i2} & \dots & A_{in} \\ \dots & \dots & \dots & \dots \\ A_{i1} & A_{i2} & \dots & A_{in} \\ \dots & \dots & \dots & \dots \\ A_{n1} & A_{n2} & \dots & A_{nn} \end{vmatrix} \tag{3-17}$$

But the determinant D_2 has two identical rows, hence its value is zero. Of course, a similar proof can be made for the columns of the determinant. Thus, it can be stated as a general rule that if the elements of a row (or column) of a determinant are multiplied by cofactors of the corresponding elements in a different row (or column) and the products are added, the result is zero.

An important property to be used later in expanding a determinant by the method of pivotal condensation is the following. The value of a determinant is unchanged if a scalar multiple of one row (or column) is added to another row (or column). This property can be demonstrated by again referring to the third-order determinant $|\mathbf{A}|$ (see Equation 3-15). As an example, assume that γ times the elements of the second row of $\mathbf{A}$ are added to the elements of the first row, giving a new determinant:

$$\begin{vmatrix} A_{11} + \gamma A_{21} & A_{12} + \gamma A_{22} & A_{13} + \gamma A_{23} \\ A_{21} & A_{22} & A_{23} \\ A_{31} & A_{32} & A_{33} \end{vmatrix}$$

When expanded by cofactors of the elements of the first row, the value of this determinant is

$$(A_{11} + \gamma A_{21})A_{11}^c + (A_{12} + \gamma A_{22})A_{12}^c + (A_{13} + \gamma A_{23})A_{13}^c \tag{c}$$

in which A_{11}^c, A_{12}^c, and A_{13}^c are the cofactors of elements A_{11}, A_{12}, and A_{13} in the original determinant $|\mathbf{A}|$. When expanded further, expression (c) becomes

$$A_{11}A_{11}^c + A_{12}A_{12}^c + A_{13}A_{13}^c + \lambda A_{21}A_{11}^c + \lambda A_{22}A_{12}^c + \lambda A_{23}A_{13}^c \tag{d}$$

The first three terms in this expression are seen to represent the expansion of the original determinant; that is, they are equal to $|\mathbf{A}|$. The last three terms can be expressed as a determinant in which γA_{21}, γA_{22}, and γA_{23} have taken the place of the elements A_{11}, A_{12}, and A_{13} in the original determinant. Thus, the last three terms are equal to

$$\begin{vmatrix} \gamma A_{21} & \gamma A_{22} & \gamma A_{23} \\ A_{21} & A_{22} & A_{23} \\ A_{31} & A_{32} & A_{33} \end{vmatrix}$$

However, this determinant is one in which the first row is a scalar multiple of the second row, and therefore it is equal to zero. This result means that the value of expression (d) is the same as that of the determinant $|\mathbf{A}|$. This demonstration can be generalized without difficulty, and hence it can be concuded that adding a scalar multiple of one row to another row does not change the value of the determinant. Of course, this conclusion applies to columns as well as to rows.

The preceding rule leads to a very useful scheme for evaluating a determinant. By suitable combinations of rows or columns it is possible to produce zeros in all elements but one of a selected row or column, thus simplifying the expansion of the determinant by cofactors of the elements in that row or column. This idea is the basis of the method of pivotal condensation, described in the next section.

Summary of Properties of Determinants. The most important properties discussed previously in this chapter are listed here for convenient reference:

1. The value of a determinant may be found by multiplying every element in any row (or column) by its cofactor and summing the products.
2. If any two rows (or columns) of a determinant are interchanged, the value of the determinant will change sign.
3. The determinant of any square matrix is equal to the determinant of its transpose.
4. If all elements of one row (or column) of a determinant are multiplied by a scalar λ, the value of the determinant is also multiplied by λ. If

all elements of j rows (or columns) of a determinant are multiplied by λ, the value of the determinant is multiplied by λ^j.

5. If a square matrix of order n is multiplied by a scalar λ, its determinant is multiplied by λ^n.
6. If two square matrices of the same order are multiplied, the determinant of their product is equal to the product of their determinants. In general, the determinant of the product of any number of square matrices of the same order is equal to the product of their determinants.
7. If the product of two or more square matrices yields the null matrix, then at least one of the matrices is singular (that is, its determinant is zero).
8. If a rectangular matrix of order $m \times n$ is postmultiplied by a rectangular matrix of order $n \times m$, and if $m > n$, then the product is a singular matrix of order m.
9. The determinant of a diagonal matrix or a triangular matrix is equal to the product of the diagonal elements.
10. A determinant will have the value zero whenever any of the following conditions is met:
    (a) All elements of a row (or column) are zero.
    (b) All cofactors of the elements of a row (or column) are zero.
    (c) Two rows (or columns) are identical.
    (d) All elements of one row (or column) are scalar multiples of the corresponding elements of another row (or column).
    (e) The rows (or columns) are linearly dependent; that is, one row (or column) is a linear combination of the other rows (or columns).
11. If the elements of one row (or column) of a determinant are multiplied by cofactors of the corresponding elements of a different row (or column) and the products are added, the result is always zero.
12. The value of a determinant is unchanged if a scalar multiple of one row (or column) is added to another row (or column). This rule can be used to produce zeros in all elements but one of a selected row or column.

3.4. Evaluation by Pivotal Condensation. The method discussed in this section for the numerical evaluation of a determinant makes use of the last property described in the preceding section. The procedure is to reduce to zero all but one element in a particular row or column. Then the determinant can be evaluated in terms of the remaining element and its cofactor. The cofactor will itself contain a determinant of lower order, and hence the process must be repeated. Eventually the determinant will be reduced to order three, whereupon the remaining evaluation can be performed without difficulty. The first few steps in the reduction are illustrated by the following example in which it is assumed that a determinant of order n is to be evaluated:

$$|\mathbf{A}| = \begin{vmatrix} A_{11} & A_{12} & A_{13} & \dots & A_{1n} \\ A_{21} & A_{22} & A_{23} & \dots & A_{2n} \\ A_{31} & A_{32} & A_{33} & \dots & A_{3n} \\ \dots & \dots & \dots & \dots & \dots \\ A_{n1} & A_{n2} & A_{n3} & \dots & A_{nn} \end{vmatrix}$$

Suppose that the second row is selected for the expansion and that all elements in that row are to be reduced to zero except A_{22}. The element A_{22} is known as the *pivotal element*, or simply the *pivot*. The first step in the expansion is to reduce the pivot element A_{22} to unity, either by dividing all elements of the second row or all elements of the second column by A_{22}. If the second row is divided by the pivot element, the following determinant is obtained:

$$|\mathbf{A}| = A_{22} \begin{vmatrix} A_{11} & A_{12} & A_{13} & \dots & A_{1n} \\ \dfrac{A_{21}}{A_{22}} & 1 & \dfrac{A_{23}}{A_{22}} & \dots & \dfrac{A_{2n}}{A_{22}} \\ A_{31} & A_{32} & A_{33} & \dots & A_{3n} \\ \dots & \dots & \dots & \dots & \dots \\ A_{n1} & A_{n2} & A_{n3} & \dots & A_{nn} \end{vmatrix}$$

The next step is to add scalar multiples of the second column to the other columns so as to make all elements of the second row except the pivot element equal to zero. For example, the ratio $-A_{21}/A_{22}$ times the second column, when added to the first column, will produce the desired zero element in the first column:

$$|\mathbf{A}| = A_{22} \begin{vmatrix} A_{11} - \dfrac{A_{21}}{A_{22}} A_{12} & A_{12} & A_{13} & \dots & A_{1n} \\ 0 & 1 & \dfrac{A_{23}}{A_{22}} & \dots & \dfrac{A_{2n}}{A_{22}} \\ A_{31} - \dfrac{A_{21}}{A_{22}} A_{32} & A_{32} & A_{33} & \dots & A_{3n} \\ \dots & \dots & \dots & \dots & \dots \\ A_{n1} - \dfrac{A_{21}}{A_{22}} A_{n2} & A_{n2} & A_{n3} & \dots & A_{nn} \end{vmatrix}$$

The same operation can be performed with columns 3, 4, . . ., n, resulting in an expression of the form

$$|\mathbf{A}| = A_{22} \begin{vmatrix} B_{11} & B_{12} & B_{13} & \dots & B_{1n} \\ 0 & 1 & 0 & \dots & 0 \\ B_{31} & B_{32} & B_{33} & \dots & B_{3n} \\ \dots & \dots & \dots & \dots & \dots \\ B_{n1} & B_{n2} & B_{n3} & \dots & B_{nn} \end{vmatrix}$$

The determinant in this expression is equal to the pivot element times its cofactor, which is a determinant of order $n - 1$ consisting of the elements B_{11}, B_{13}, . . ., B_{nn}. By continuing this procedure the determinant is reduced successively to determinants of smaller order. Eventually a determinant of order three is obtained, and the evaluation can be completed without further difficulty. Because of the use of a pivot element with each successive condensation of the determinant, this technique is known as the method of *pivotal condensation*.

As a numerical example using the method of pivotal condensation, assume that the following determinant D is to be evaluated:

$$D = \begin{vmatrix} 2 & 3 & 4 & 2 \\ 10 & 3 & -2 & 0 \\ 1 & 2 & 4 & 3 \\ -2 & 0 & 1 & 5 \end{vmatrix}$$

Since zero elements already appear in the determinant, it is natural to expand the determinant by a row or column containing zero elements; assume that the last column is selected. Next, a pivot is selected in the last column; assume that the first element in that column is selected. Then either the first row or the last column must be divided by two, which is the value of the pivot element. If the first row is divided by two, the determinant takes the form:

$$D = 2 \begin{vmatrix} 1 & 1.5 & 2 & 1 \\ 10 & 3 & -2 & 0 \\ 1 & 2 & 4 & 3 \\ -2 & 0 & 1 & 5 \end{vmatrix}$$

Next, three times the first row is subtracted from the third row, and five times the first row is subtracted from the last row:

$$D = 2 \begin{vmatrix} 1 & 1.5 & 2 & 1 \\ 10 & 3 & -2 & 0 \\ -2 & -2.5 & -2 & 0 \\ -7 & -7.5 & -9 & 0 \end{vmatrix}$$

Now the determinant can be expanded by elements of the last column to yield

$$D = (2)(-1)\begin{vmatrix} 10 & 3 & -2 \\ -2 & -2.5 & -2 \\ -7 & -7.5 & -9 \end{vmatrix}$$

The evaluation of the third-order determinant can be carried out either by pivotal condensation or by following the rule portrayed in Figure 3-1b; the final result is

$$D = (2)(-1)(68) = -136$$

Pivotal Condensation Using Largest Pivots. Although it is true that in theory any element may be selected as a pivot, as was implied in the preceding examples, there may be practical difficulties if a suitable choice of the pivot element is not made. Errors arise in the calculations from rounding off the numbers, and it is usually desirable to minimize this loss of accuracy. This result can be accomplished by using the largest possible pivot elements. One method is to select the numerically largest element in the determinant as the first pivot and then to proceed with the reduction of the determinant to one of lower order. The process is repeated with the largest element of the new determinant selected as the second pivot; this technique is continued until the determinant is evaluated.

A more commonly used method is to interchange rows and columns in the determinant so that the largest element appears in the leading position at the upper left-hand corner of the determinant. Then this element is used as the pivot element in the process of condensation, and the order of the determinant is reduced by one. Next, the entire process is repeated for the new determinant, with rows and columns being interchanged until the largest element is again in the leading position. The same steps are followed for each new determinant until eventually the value of the determinant is obtained. Of course, it is necessary to take account of the number of row and column interchanges, since each interchange means a change in the sign of the determinant.

To illustrate the evaluation of a determinant by this systematic means, the example given previously will be considered again. In the present example, the evaluation will be performed with operations on rows only, always using the largest pivots. Of course, a similar procedure could be followed by operating on columns. The given determinant D is

$$D = \begin{vmatrix} 2 & 3 & 4 & 2 \\ 10 & 3 & -2 & 0 \\ 1 & 2 & 4 & 3 \\ -2 & 0 & 1 & 5 \end{vmatrix}$$

Since the largest element (ten) appears in the second row, the first two rows will be interchanged so that the pivot appears in the leading position. Also, the first row is divided by the pivot element:

$$D=-10\begin{vmatrix} 1 & 0.3 & -0.2 & 0 \\ 2 & 3 & 4 & 2 \\ 1 & 2 & 4 & 3 \\ -2 & 0 & 1 & 5 \end{vmatrix}$$

The minus sign is introduced before the determinant because of the interchange of rows. Multiples of the first row are now added to the remaining rows to make all elements in the first column equal to zero except for the pivot element. Thus, the determinant becomes

$$D=-10\begin{vmatrix} 1 & 0.3 & -0.2 & 0 \\ 0 & 2.4 & 4.4 & 2 \\ 0 & 1.7 & 4.2 & 3 \\ 0 & 0.6 & 0.6 & 5 \end{vmatrix}$$

or

$$D=-10(1)\begin{vmatrix} 2.4 & 4.4 & 2 \\ 1.7 & 4.2 & 3 \\ 0.6 & 0.6 & 5 \end{vmatrix}$$

The largest element remaining in the determinant is five, which can be brought to the leading position by interchanging the first and third columns and the first and third rows. Then, after dividing the first row by five, the determinant is

$$D=-10(5)\begin{vmatrix} 1 & 0.12 & 0.12 \\ 3 & 4.2 & 1.7 \\ 2 & 4.4 & 2.4 \end{vmatrix}$$

Since there were two interchanges of rows and columns, no change in sign before the determinant is needed in this case. Next, appropriate multiples of the first row are added to the second and third rows to give

$$D=-50\begin{vmatrix} 1 & 0.12 & 0.12 \\ 0 & 3.84 & 1.34 \\ 0 & 4.16 & 2.16 \end{vmatrix}$$

from which

$$D=-50\begin{vmatrix} 3.84 & 1.34 \\ 4.16 & 2.16 \end{vmatrix}=-136$$

which is the same result as before.

In addition to pivotal condensation, there are many other methods that have been developed for evaluating determinants; such methods can be found in the references listed at the end of the book.

Problems

3.2-1. Expand the third-order determinant given in Equation (3-2) by cofactors of elements of the second column, and show that the result is the same as the one obtained from Equation (3-3).

3.2-2. Evaluate the determinant D_1 shown below by (a) using the expansion rule shown in Figure 3-1b, (b) using cofactors of the elements in a row, and (c) using cofactors of the elements in a column.

$$D_1 = \begin{vmatrix} 3 & 1 & -2 \\ -1 & 4 & -5 \\ 4 & 0 & 1 \end{vmatrix}$$

3.2-3, Repeat the preceding problem for the following determinant:

$$D_2 = \begin{vmatrix} -2 & 1 & 2 \\ 4 & 0 & 1 \\ -3 & 5 & 2 \end{vmatrix}$$

3.2-4. Repeat Problem 3.2-2 for the following determinant:

$$D_3 = \begin{vmatrix} x & -x & 4x \\ y & 2y & 0 \\ -z & 3z & 2z \end{vmatrix}$$

3.2-5. For what value of x will the following determinant be equal to zero?

$$D_4 = \begin{vmatrix} 3 & 1 & 2 \\ -6 & 4 & x \\ -2 & 0 & 1 \end{vmatrix}$$

3.2-6. For what values of a and b will the cofactor of each element in the following determinant be the same as the element?

$$D_5 = \begin{vmatrix} a & b & b \\ -b & -a & b \\ -b & b & -a \end{vmatrix}$$

3.2-7. Evaluate the following fourth-order determinant using the method of expansion by cofactors:

$$D_6 = \begin{vmatrix} 2 & 1 & 0 & -3 \\ 1 & -2 & 4 & 5 \\ 3 & 0 & 1 & 4 \\ -2 & 0 & 3 & 1 \end{vmatrix}$$

3.2-8. Evaluate the following determinant using the method of expansion by cofactors:

$$D_7 = \begin{vmatrix} 5 & -15 & 0 & 2 \\ 0 & 1 & 5 & 10 \\ -1 & 5 & -5 & 0 \\ 2 & 0 & 1 & 10 \end{vmatrix}$$

3.2-9. Obtain a formula for the value of the determinant of an upper triangular matrix **U**:

$$|\mathbf{U}| = \begin{vmatrix} A_{11} & A_{12} & A_{13} & \dots & A_{1n} \\ 0 & A_{22} & A_{23} & \dots & A_{2n} \\ 0 & 0 & A_{33} & \dots & A_{3n} \\ \dots & \dots & \dots & \dots & \dots \\ 0 & 0 & 0 & \dots & A_{nn} \end{vmatrix}$$

3.2-10. The following equation is valid for a tridiagonal determinant of order n as shown:

$$\begin{vmatrix} 2 & -1 & 0 & 0 & \dots & 0 & 0 & 0 \\ -1 & 2 & -1 & 0 & \dots & 0 & 0 & 0 \\ 0 & -1 & 2 & -1 & \dots & 0 & 0 & 0 \\ \dots & \dots & \dots & \dots & \dots & \dots & \dots & \dots \\ 0 & 0 & 0 & 0 & \dots & -1 & 2 & -1 \\ 0 & 0 & 0 & 0 & \dots & 0 & -1 & 2 \end{vmatrix} = n + 1$$

Demonstrate the validity of this equation for $n = 1, 2, 3,$ and 4 by expanding by cofactors.

3.2-11. A triangle abc in the xy plane has vertices with coordinates (x_a, y_a), (x_b, y_b), and (x_c, y_c). The vertices are in the order abc when going counterclockwise around the triangle. Show that the area A_{abc} of the triangle is given by the following expression:

$$A_{abc} = \frac{1}{2}\begin{vmatrix} 1 & 1 & 1 \\ x_a & x_b & x_c \\ y_a & y_b & y_c \end{vmatrix}$$

3.3–1. Show that the value of the determinant D_6 in Problem 3.2–7 changes sign when rows 1 and 3 are interchanged.

3.3–2. Show that the value of the determinant D_7 in Problem 3.2–8 changes sign when columns 1 and 4 are interchanged.

3.3–3. Using the second-order matrices **A** and **B** shown below, demonstrate that the determinant of their product is equal to the product of their determinants (see Equation 3–7).

$$\mathbf{A} = \begin{bmatrix} A_{11} & A_{12} \\ A_{21} & A_{22} \end{bmatrix} \qquad \mathbf{B} = \begin{bmatrix} B_{11} & B_{12} \\ B_{21} & B_{22} \end{bmatrix}$$

3.3–4. Using the matrices **C** and **D** shown below, demonstrate that the determinant of their product is equal to the product of their determinants.

$$\mathbf{C} = \begin{bmatrix} 2 & -1 & 3 \\ -4 & 1 & 0 \\ 0 & -5 & 6 \end{bmatrix} \qquad \mathbf{D} = \begin{bmatrix} 1 & 7 & 0 \\ 3 & 2 & 1 \\ 4 & 0 & 2 \end{bmatrix}$$

3.3–5. Using the matrices **E**, **F**, and **G** shown below, demonstrate that the determinant of their product is equal to the product of their determinants.

$$\mathbf{E} = \begin{bmatrix} 2 & 7 & 1 \\ 0 & 2 & -5 \\ 0 & 0 & 1 \end{bmatrix} \qquad \mathbf{F} = \begin{bmatrix} -2 & -1 & 0 \\ 3 & 2 & 0 \\ -4 & 5 & 3 \end{bmatrix} \qquad \mathbf{G} = \begin{bmatrix} 0 & 6 & 4 \\ 2 & 0 & 2 \\ -4 & 5 & 0 \end{bmatrix}$$

3.3–6. Show that the product **AB** of the following rectangular matrices is a singular matrix (see Equation 3–10):

$$\mathbf{A} = \begin{bmatrix} 6 & -3 \\ 1 & 4 \\ -2 & 1 \end{bmatrix} \qquad \mathbf{B} = \begin{bmatrix} 2 & -1 & -2 \\ 3 & -4 & -1 \end{bmatrix}$$

3.3–7. Using the matrices **C** and **D** of Problem 3.3–4, calculate the determinant of the sum $\mathbf{C} + \mathbf{D}$, compare it with the sum of the determinants, and observe that $|\mathbf{C} + \mathbf{D}| \neq |\mathbf{C}| + |\mathbf{D}|$.

3.3–8. Demonstrate the rule given by Equation (3–14) with the following matrix:

$$\mathbf{D} = \begin{bmatrix} a_{11} & a_{12} & a_{13} & 0 & 0 \\ a_{21} & a_{22} & a_{23} & 0 & 0 \\ a_{31} & a_{32} & a_{33} & 0 & 0 \\ b_{11} & b_{12} & b_{13} & c_{11} & c_{12} \\ b_{21} & b_{22} & b_{23} & c_{21} & c_{22} \end{bmatrix}$$

(Hint: Expand $|\mathbf{D}|$ by cofactors of the elements in the last column.)

3.3–9. Show that the following equation is valid for determinants of any order n:

$$D_1 = \begin{vmatrix} a & b & b & \dots & b \\ b & a & b & \dots & b \\ \dots & \dots & \dots & \dots & \dots \\ b & b & b & \dots & a \end{vmatrix} = [a + b(n-1)]\,(a-b)^{n-1}$$

(Hint: Subtract the last row from each of the other rows; add columns 1 to $n-1$ to the last column; then expand by cofactors of the elements of the last column.)

3.3–10. Show that the following equation is valid for determinants of any order n:

$$D_2 = \begin{vmatrix} 1+a_1 & a_2 & \dots & a_n \\ a_1 & 1+a_2 & \dots & a_n \\ \dots & \dots & \dots & \dots \\ a_1 & a_2 & \dots & 1+a_n \end{vmatrix} = 1 + a_1 + a_2 + \dots + a_n$$

(Hint: Follow the steps described for the preceding problem.)

3.4–1. Using the method of pivotal condensation, evaluate the determinant D_1 shown below.

$$D_1 = \begin{vmatrix} 2 & 0 & 1 & -1 \\ 3 & -2 & 1 & 5 \\ -3 & 0 & 2 & 1 \\ 0 & 1 & 1 & -2 \end{vmatrix}$$

3.4–2. Evaluate the determinant D_6 in Problem 3.2–7 using pivotal condensation.

3.4–3. Evaluate the determinant D_7 in Problem 3.2–8 using pivotal condensation.

3.4–4. Evaluate the fourth-order determinant D given in Section 3.2 by using pivotal condensation.

3.4–5. Evaluate the following determinant using pivotal condensation:

$$D_2 = \begin{vmatrix} 4 & -2 & 1 & 3 & 0 \\ 5 & 0 & 5 & 4 & 1 \\ -1 & 3 & 2 & 8 & -2 \\ 2 & -4 & 0 & 3 & 4 \\ -3 & 1 & 6 & -1 & -2 \end{vmatrix}$$

3.4-6. Evaluate the following determinant using pivotal condensation:

$$D_3 = \begin{vmatrix} 1 & -1 & 2 & 4 & 0 & -3 \\ -2 & 0 & -1 & 3 & 1 & -2 \\ 1 & 2 & 0 & -3 & 1 & 4 \\ 3 & -1 & 1 & -2 & 0 & 4 \\ 0 & -2 & 2 & 2 & -3 & 2 \\ 2 & 0 & -1 & 3 & 1 & 1 \end{vmatrix}$$

4

...Inverse of a Matrix...

In Chapter 2 the matrix operations of addition, subtraction, and multiplication were described. The matrix operation which is analogous to division is known as *inversion*, and methods for calculating the *inverse* of a matrix are presented in this chapter.

4.1. Definitions. The inverse of a square matrix **A** is written as $\mathbf{A}^{-1}$, and it is defined as the matrix which, when multiplied by the original matrix **A**, results in the identity matrix. The inverse is always a square matrix of the same order as the original matrix itself, and only a square matrix has an inverse. Thus, the relationship between a matrix and its inverse is given by the expression

$$\mathbf{A}\mathbf{A}^{-1} = \mathbf{A}^{-1}\mathbf{A} = \mathbf{I} \qquad (4\text{-}1)$$

Equation (4-1) shows that a matrix and its inverse are commutative.

There is an analogy between the reciprocal of a number in scalar algebra and the inverse of a matrix. In the former case, two numbers a and b may be related by the expression $ab = 1$, and therefore b equals a^{-1}. In matrix algebra, the corresponding relations are

$$\mathbf{A}\mathbf{B} = \mathbf{I} \qquad (4\text{-}2)$$

from which it follows that **B** is the inverse of **A**, or

$$\mathbf{B} = \mathbf{A}^{-1} \qquad (4\text{-}3a)$$

Of course, it is also true that **A** is the inverse of **B**; that is,

$$\mathbf{A} = \mathbf{B}^{-1} \qquad (4\text{-}3b)$$

To illustrate the use of the matrix inverse when solving simultaneous equations, consider a set of equations of the form

$$\mathbf{A}\mathbf{X} = \mathbf{D} \qquad (4\text{-}4)$$

in which **A** is a square matrix of coefficients, **X** is a column matrix of the unknowns, and **D** is a column matrix of constant terms. If there are n simultaneous equations, the order of the matrix **A** wil be $n \times n$, and the order of both **X** and **D** will be $n \times 1$. The elements of the matrices **A** and **D** are assumed to be known quantities. If both sides of Equation (4–4) are premultiplied by $\mathbf{A}^{-1}$, the result is

$$\mathbf{A}^{-1}\mathbf{AX} = \mathbf{A}^{-1}\mathbf{D}$$

from which is obtained (see Equation 4–1)

$$\mathbf{IX} = \mathbf{A}^{-1}\mathbf{D}$$

Since the multiplication of a matrix by the identity matrix gives the same matrix, the preceding equation becomes

$$\mathbf{X} = \mathbf{A}^{-1}\mathbf{D} \tag{4-5}$$

This last expression gives a convenient way of representing the solution of a set of simultaneous equations, and it shows also that finding the inverse of a matrix is essentially the same as solving the equations.

4.2. Cofactor Matrix. In the development of a formal method for obtaining the inverse of a matrix, it is desirable to define a *cofactor matrix*. A cofactor matrix is constructed by replacing each element of a square matrix by its cofactor. The cofactors are obtained from the determinant of the original matrix, as explained in Chapter 3. To illustrate the construction of a cofactor matrix, consider the third-order matrix **A**:

$$\mathbf{A} = \begin{bmatrix} A_{11} & A_{12} & A_{13} \\ A_{21} & A_{22} & A_{23} \\ A_{31} & A_{32} & A_{33} \end{bmatrix} \tag{4-6}$$

The cofactor matrix obtained from **A** is denoted $\mathbf{A}^c$ and appears as follows:

$$\mathbf{A}^c = \begin{bmatrix} A_{11}^c & A_{12}^c & A_{13}^c \\ A_{21}^c & A_{22}^c & A_{23}^c \\ A_{31}^c & A_{32}^c & A_{33}^c \end{bmatrix} \tag{4-7}$$

Each of the cofactors in this matrix is obtained from the determinant of **A**, which is

$$|\mathbf{A}| = \begin{vmatrix} A_{11} & A_{12} & A_{13} \\ A_{21} & A_{22} & A_{23} \\ A_{31} & A_{32} & A_{33} \end{vmatrix}$$

Thus, the cofactors are

$$A_{11}^c = \begin{vmatrix} A_{22} & A_{23} \\ A_{32} & A_{33} \end{vmatrix} \qquad A_{12}^c = -\begin{vmatrix} A_{21} & A_{23} \\ A_{31} & A_{33} \end{vmatrix}$$

and so forth for the remaining cofactors. From this example of a 3×3 matrix, it is evident that a similar method can be used to obtain the cofactor matrix for a square matrix of any order. In all cases, the cofactor matrix has the same order as the original matrix.

As a numerical example, consider the following square matrix **A**, and assume that its cofactor matrix is to be obtained:

$$\mathbf{A} = \begin{bmatrix} 2 & -1 & 3 \\ 0 & 4 & -2 \\ 1 & -3 & 5 \end{bmatrix}$$

The elements of the first row of the cofactor matrix are determined as follows:

$$A_{11}^c = \begin{vmatrix} 4 & -2 \\ -3 & 5 \end{vmatrix} = 14$$

$$A_{12}^c = -\begin{vmatrix} 0 & -2 \\ 1 & 5 \end{vmatrix} = -2$$

$$A_{13}^c = \begin{vmatrix} 0 & 4 \\ 1 & -3 \end{vmatrix} = -4$$

By continuing in this manner, all elements can be determined readily, and the cofactor matrix becomes

$$\mathbf{A}^c = \begin{bmatrix} 14 & -2 & -4 \\ -4 & 7 & 5 \\ -10 & 4 & 8 \end{bmatrix}$$

From the definition of a cofactor matrix it can be seen that if the original matrix **A** is symmetric, the cofactor matrix also will be symmetric. Similarly, if **A** is a diagonal matrix, then the cofactor matrix is a diagonal matrix; and if **A** is a lower triangular matrix, the cofactor matrix is upper triangular, and vice versa.

4.3. Adjoint Matrix. Another matrix to be defined is the *adjoint matrix* (also called the *adjugate matrix*), which is simply the transpose of the cofactor matrix. Thus, if the illustration of the preceding article is continued (see Equations 4-6 and 4-7), the adjoint of **A**, denoted as $\mathbf{A}^a$, will be given by the expression

$$\mathbf{A}^a = (\mathbf{A}^c)^{\mathbf{T}} = \begin{bmatrix} A_{11}^c & A_{21}^c & A_{31}^c \\ A_{12}^c & A_{22}^c & A_{32}^c \\ A_{13}^c & A_{23}^c & A_{33}^c \end{bmatrix} \tag{4-8}$$

Specifically, the adjoint matrix for the numerical example of the preceding section is

$$\mathbf{A}^a = \begin{bmatrix} 14 & -4 & -10 \\ -2 & 7 & 4 \\ -4 & 5 & 8 \end{bmatrix}$$

Inasmuch as the adjoint matrix is the transpose of the cofactor matrix, it can be concluded that if **A** is symmetric, $\mathbf{A}^a$ is symmetric, and if **A** is diagonal, $\mathbf{A}^a$ is diagonal. In the case of a triangular matrix, the adjoint matrix is of the same form as the original matrix (that is, if **A** is lower triangular, $\mathbf{A}^a$ is also lower triangular).

4.4. Inverse of a Matrix. In order to show how the adjoint matrix is related to the operation of inversion, it is necessary to observe the result of multiplying a matrix by its adjoint. For the third-order matrix **A** in the above example, the multiplication is as follows:

$$\mathbf{A}\mathbf{A}^a = \begin{bmatrix} A_{11} & A_{12} & A_{13} \\ A_{21} & A_{22} & A_{23} \\ A_{31} & A_{32} & A_{33} \end{bmatrix} \begin{bmatrix} A_{11}^c & A_{21}^c & A_{31}^c \\ A_{12}^c & A_{22}^c & A_{32}^c \\ A_{13}^c & A_{23}^c & A_{33}^c \end{bmatrix}$$

Now suppose that the product matrix is denoted as **B**:

$$\mathbf{A}\mathbf{A}^a = \mathbf{B} = \begin{bmatrix} B_{11} & B_{12} & B_{13} \\ B_{21} & B_{22} & B_{23} \\ B_{31} & B_{32} & B_{33} \end{bmatrix}$$

It follows that the elements of **B** are obtained as the inner products of the rows of **A** and the columns of $\mathbf{A}^a$. For example, the element B_{11} is equal to the inner product of the first row of **A** and the first column of $\mathbf{A}^a$:

$$B_{11} = A_{11}A_{11}^c + A_{12}A_{12}^c + A_{13}A_{13}^c$$

The other elements of **B** are obtained in a similar manner:

$$B_{12} = A_{11}A_{21}^c + A_{12}A_{22}^c + A_{13}A_{23}^c$$
$$B_{13} = A_{11}A_{31}^c + A_{12}A_{32}^c + A_{13}A_{33}^c$$
$$B_{21} = A_{21}A_{11}^c + A_{22}A_{12}^c + A_{23}A_{13}^c$$
$$B_{22} = A_{21}A_{21}^c + A_{22}A_{22}^c + A_{23}A_{23}^c$$

and so forth until all nine elements are calculated. Inspection of the expression for B_{11} shows that it consists of the elements in the first row of **A** times the cofactors of the same elements. Thus, B_{11} is identical to the expansion of $|\mathbf{A}|$ by cofactors of the elements of the first row, and therefore its value must be the same as the value of the determinant $|\mathbf{A}|$. The expression for B_{12} also contains the elements of the first row of **A**, but in this case they are multiplied by cofactors of the second row of $|\mathbf{A}|$. Such an expansion produces a value of zero, as shown in Section 3.3. As the remaining expressions for the elements B_{13}, B_{21}, etc., are examined, it is found that all are zero except those on the principal diagonal, that is, B_{11}, B_{22}, and B_{33}, and each of those elements is equal to $|\mathbf{A}|$. Thus, it is concluded that the result of multiplying the matrix **A** by its adjoint $\mathbf{A}^a$ is

$$\mathbf{AA}^a = \begin{bmatrix} |\mathbf{A}| & 0 & 0 \\ 0 & |\mathbf{A}| & 0 \\ 0 & 0 & |\mathbf{A}| \end{bmatrix}$$

This expression can be simplified by introducing the identity matrix:

$$\mathbf{AA}^a = |\mathbf{A}|\mathbf{I}$$

Dividing both sides of this equation by the determinant of **A** gives

$$\mathbf{A}\frac{\mathbf{A}^a}{|\mathbf{A}|} = \mathbf{I}$$

from which it is seen that (compare with Equation 4–1):

$$\mathbf{A}^{-1} = \frac{\mathbf{A}^a}{|\mathbf{A}|} \tag{4-9}$$

This equation states that the inverse of a matrix is equal to the adjoint matrix divided by the determinant of the matrix. Thus, a formal means of obtaining the inverse of a matrix has been found. The result applies to a square matrix of any order, since the preceding development for a third-order matrix could have been carried out in more general terms for a matrix of order n.

Equation (4–9) shows that the inverse of a matrix exists only if the determinant of the matrix is not zero. If the determinant is equal to zero, that is, if

$$|\mathbf{A}| = 0 \tag{4-10}$$

the matrix **A** is said to be *singular* and no inverse can be found. From the previous discussion of properties of determinants (see Section 3.3) it can be seen that there are several conditions which will cause a matrix to be singular. To be specific, a matrix is singular if all elements of one row are zero, or if all cofactors of the elements of one row are zero, or if two rows are the same, or if one row is a scalar multiple of another row, or (in general) if one row is a linear combination of other rows. Of course, all of these conditions apply to columns of the matrix as well as to the rows. In most cases, it is not possible to identify a singular matrix merely by inspection of the matrix. Therefore, the usual test for singularity is to evaluate the determinant of the matrix in order to determine if it is equal to zero.

In Chapter 2, it was observed that if **AB** = **AC**, it does not necessarily follow that **B** = **C**. In the case of square matrices, it may now be noted that if **A** is singular, **B** and **C** are not necessarily equal (see Example 6 of Section 2.4). However, if **A** is nonsingular, then **B** must equal **C** (even when **B** and **C** are singular).

In order to illustrate the calculation of a matrix inverse by using Equation (4–9), two examples will be solved. Assume first that it is desired to calculate the inverse of the second-order matrix

$$\mathbf{A} = \begin{bmatrix} 2 & -1 \\ 3 & -4 \end{bmatrix}$$

The first step is to construct the cofactor matrix, which is

$$\mathbf{A}^c = \begin{bmatrix} -4 & -3 \\ 1 & 2 \end{bmatrix}$$

and, therefore, the adjoint matrix is

$$\mathbf{A}^c = \begin{bmatrix} -4 & 1 \\ -3 & 2 \end{bmatrix}$$

Next, the determinant of **A** is calculated:

$$|\mathbf{A}| = -8 + 3 = -5$$

Finally, the inverse of **A** is found by dividing the adjoint matrix by the value of the determinant:

$$\mathbf{A}^{-1} = \frac{\mathbf{A}^a}{|\mathbf{A}|} = \begin{bmatrix} 0.8 & -0.2 \\ 0.6 & -0.4 \end{bmatrix}$$

This result for the inverse can be checked by multiplying **A** and $\mathbf{A}^{-1}$:

$$\mathbf{A}\mathbf{A}^{-1} = \begin{bmatrix} 2 & -1 \\ 3 & -4 \end{bmatrix} \begin{bmatrix} 0.8 & -0.2 \\ 0.6 & -0.4 \end{bmatrix} = \begin{bmatrix} 1 & 0 \\ 0 & 1 \end{bmatrix}$$

Since the result is the identity matrix, the inverse has been correctly calculated. The identity matrix is obtained also if the product $\mathbf{A}^{-1}\mathbf{A}$ is calculated.

As a second example, assume that the following matrix of order three is to be inverted:

$$\mathbf{B} = \begin{bmatrix} 3 & 2 & 0 \\ -1 & -2 & 4 \\ 2 & -1 & -3 \end{bmatrix}$$

As in the previous example, the first step is to calculate the cofactor matrix:

$$\mathbf{B}^c = \begin{bmatrix} 10 & 5 & 5 \\ 6 & -9 & 7 \\ 8 & -12 & -4 \end{bmatrix}$$

and from it the adjoint matrix is obtained:

$$\mathbf{B}^a = \begin{bmatrix} 10 & 6 & 8 \\ 5 & -9 & -12 \\ 5 & 7 & -4 \end{bmatrix}$$

The determinant of **B** is equal to 40; therefore, the inverse is (see Equation 4-9)

$$\mathbf{B}^{-1} = \frac{1}{40}\begin{bmatrix} 10 & 6 & 8 \\ 5 & -9 & -12 \\ 5 & 7 & -4 \end{bmatrix}$$

This result can be checked readily by multiplying $\mathbf{B}^{-1}$ and **B**.

Once the preceding method is understood, the inverse of a second-order matrix can be written by inspection. Let the following matrix **A** represent any nonsingular second-order matrix:

$$\mathbf{A} = \begin{bmatrix} A_{11} & A_{12} \\ A_{21} & A_{22} \end{bmatrix} \tag{4-11}$$

Then, after finding the cofactor and adjoint matrices, it can be seen that the following expression gives the inverse of **A**:

$$\mathbf{A}^{-1} = \frac{1}{D}\begin{bmatrix} A_{22} & -A_{12} \\ -A_{21} & A_{11} \end{bmatrix} \tag{4-12}$$

in which D is the determinant of the matrix, that is, $D = |\mathbf{A}|$. Thus, the following rule holds: The inverse of a 2 × 2 matrix is found by interchanging the elements on the principal diagonal, changing the signs of the other two elements, and then dividing by the value of the determinant.

The method of inverting a matrix by using the adjoint matrix is not practical for large matrices. Instead, a more systematic procedure is desirable, and a variety of such schemes have been developed. One of these is the method of successive transformations, which is suitable for hand calculations with matrices that are not too large; this method is described in Section 4.7. The method of inversion by partitioning, which reduces the size of the matrices to be inverted but requires that more inversions be performed, is described in Section 4.8. Descriptions of other methods can be found in the references.

4.5. Properties of the Inverse. There are many useful properties and rules pertaining to the inverse of a matrix, and a few of the more important ones will be mentioned in this section. A somewhat obvious property is that the inverse of a matrix inverse is the original matrix itself; thus:

$$(\mathbf{A}^{-1})^{-1} = \mathbf{A} \tag{4-13}$$

provided that **A** is not singular. Also, the inverse is unique, and if $\mathbf{AB} = \mathbf{I}$ and $\mathbf{AC} = \mathbf{I}$, then it follows that $\mathbf{B} = \mathbf{C} = \mathbf{A}^{-1}$. A simple way to determine if one matrix is the inverse of another is to multiply them and see if the identity matrix is obtained.

There are some matrices that are unchanged by inversion, that is, the matrix is its own inverse. It is obvious that the identity matrix is a matrix of this kind, since the following relations hold:

$$\mathbf{II} = \mathbf{I} \qquad \text{or} \qquad \mathbf{I} = \mathbf{I}^{-1}$$

Symmetric orthogonal matrices (see Section 4.6) also have this property, as does a matrix having ones on the secondary diagonal and zeroes elsewhere. Another interesting example is given in Problem 4.4-12.

The inverse of a diagonal matrix can be obtained quite easily, since it consists of another diagonal matrix having each element on the diagonal equal to the reciprocal of the corresponding element in the original matrix. Thus, if a diagonal matrix **D** is written in the form

$$\mathbf{D} = \begin{bmatrix} \lambda_1 & 0 & \cdots & 0 \\ 0 & \lambda_2 & \cdots & 0 \\ \cdots & \cdots & \cdots & \cdots \\ 0 & 0 & \cdots & \lambda_n \end{bmatrix} \tag{4-14}$$

then its inverse is

$$\mathbf{D}^{-1} = \begin{bmatrix} \dfrac{1}{\lambda_1} & 0 & \cdots & 0 \\ 0 & \dfrac{1}{\lambda_2} & \cdots & 0 \\ \cdots & \cdots & \cdots & \cdots \\ 0 & 0 & \cdots & \dfrac{1}{\lambda_n} \end{bmatrix} \tag{4-15}$$

as can be verified by multiplication of the two matrices. The inverse of a matrix with nonzero terms only on the secondary diagonal is considered in Problems 4.5-1 and 4.5-2. The inverse of a band matrix is not normally another band matrix, as illustrated in Problem 4.4-10.

Another relationship that should be noted is that the inverse of the product of a scalar λ times a square matrix **A** is equal to the reciprocal of the scalar times the inverse of the matrix, as determined by reference to Equation (4-9):

$$(\lambda \mathbf{A})^{-1} = \frac{(\lambda \mathbf{A})^a}{|\lambda \mathbf{A}|} = \frac{\lambda^{n-1} \mathbf{A}^a}{\lambda^n |\mathbf{A}|} = \frac{1}{\lambda} \mathbf{A}^{-1} \qquad (4\text{-}16)$$

Note that the adjoint matrix consists of determinants that are of order $n - 1$; therefore, factoring powers of λ from both the adjoint matrix and the denominator determinant produces the ratio $1/\lambda$ in Equation (4-16). This relationship is useful when inverting a matrix that (for convenience in calculation) has had a common term factored from each of its elements.

In the case of a lower triangular matrix, the inverse will also be a lower triangular matrix; an analogous rule holds for an upper triangular matrix. Also, the diagonal elements in the inverse of a triangular matrix are the reciprocals of the corresponding diagonal elements in the original matrix. These properties are illustrated in several of the problems at the end of the chapter.

If a matrix is symmetric, its inverse also will be symmetric, inasmuch as the adjoint matrix is symmetric. If the original matrix is skew-symmetric and of even order, then the inverse matrix is also skew-symmetric (see Problem 4.5-3). However, if a skew-symmetric matrix is of odd order, it is singular and no inverse exists (see Section 3.3).

Now suppose that two rows of a square matrix **A** are interchanged. As a consequence, two rows of the cofactor matrix (see Equation 4-7) will be interchanged. When the cofactor matrix is transposed to obtain the adjoint matrix (see Equation 4-8), the interchange of rows becomes an interchange of columns. Since the inverse of **A** is equal to the adjoint matrix divided by the determinant (see Equation 4-9), it is evident that interchanging two rows of the original matrix will cause a corresponding interchange of two columns of the inverse. Conversely, an interchange of two columns in the original matrix results in a corresponding interchange of the rows of the inverse. Sign changes that occur in the cofactor matrix and the determinant because of interchanging rows (or columns) cancel each other out, and there is no net change of signs in the inverse from such interchanges.

The inverse of the product of two square matrices is the product of the inverses, but in reverse order; thus,

$$(\mathbf{AB})^{-1} = \mathbf{B}^{-1}\mathbf{A}^{-1} \qquad (4\text{-}17)$$

This statement may be proved by premultiplying the product **AB** by its inverse in the form given by the right-hand side of Equation (4-17), producing the identity matrix as follows:

$$\mathbf{B}^{-1}\mathbf{A}^{-1}\mathbf{AB} = \mathbf{B}^{-1}\mathbf{IB} = \mathbf{B}^{-1}\mathbf{B} = \mathbf{I}$$

If several square matrices are involved in the product, the relationship becomes

$$(\mathbf{A}_1\mathbf{A}_2 \ldots \mathbf{A}_n)^{-1} = \mathbf{A}_n^{-1} \ldots \mathbf{A}_2^{-1}\mathbf{A}_1^{-1} \qquad (4\text{-}18)$$

The rule given by Equation (4-18) is analogous to that for the transpose of the product of several matrices (see Equation 2-19b) and is demonstrated in Problem 4.5-4.

In the case of the inverse of the transpose of a matrix, the following equation holds:

$$(\mathbf{A}^{\mathsf{T}})^{-1} = (\mathbf{A}^{-1})^{\mathsf{T}} \tag{4-19}$$

That is, the inverse of the transpose of a matrix is the same as the transpose of the inverse. In order to show this relationship, begin with Equation (4-1) and transpose it using Equation (2-19a):

$$(\mathbf{A}^{-1})^{\mathsf{T}}\mathbf{A}^{\mathsf{T}} = \mathbf{I}^{\mathsf{T}} = \mathbf{I}$$

Thus, $(\mathbf{A}^{-1})^{\mathsf{T}}$ must be equal to the inverse of $\mathbf{A}^{\mathsf{T}}$, and Equation (4-19) is established (see also Problem 4.5-5).

In general, it is not true that the inverse of the sum of two matrices is equal to the sum of the inverses; that is,

$$(\mathbf{A} + \mathbf{B})^{-1} \neq \mathbf{A}^{-1} + \mathbf{B}^{-1}$$

This inequality can be established by inspection simply by taking $\mathbf{B} = \mathbf{A}$ and using Equation (4-16).

Negative powers of a nonsingular, square matrix can be defined in a manner analogous to that for scalar quantities as shown by the following equation:

$$\mathbf{A}^{-n} = (\mathbf{A}^{-1})^{n} = (\mathbf{A}^{n})^{-1} \tag{4-20}$$

in which n is a positive integer. Thus, the result of raising a matrix to the power of minus three is calculated as follows:

$$\mathbf{A}^{-3} = (\mathbf{A}^{-1})^{3} = \mathbf{A}^{-1}\mathbf{A}^{-1}\mathbf{A}^{-1}$$

or, equivalently,

$$\mathbf{A}^{-3} = (\mathbf{A}^{3})^{-1} = (\mathbf{AAA})^{-1} = \mathbf{A}^{-1}\mathbf{A}^{-1}\mathbf{A}^{-1}$$

Other negative integer powers can be calculated in the same manner.

It was pointed out previously that the determinant of the product of two matrices is equal to the product of their determinants (see Equation 3-7). Hence, it follows that the product of the determinant of a matrix and the determinant of its inverse is always unity, or

$$|\mathbf{AA}^{-1}| = |\mathbf{A}|\,|\mathbf{A}^{-1}| = 1 \tag{4-21}$$

Thus, the determinant of a matrix is equal to the reciprocal of the determinant of its inverse.

4.6. Orthogonal Matrices. A special type of matrix encountered in eigenvalue problems and in rotation of axes (see Chapters 6 and 7) is an *orthogo-*

nal matrix. An orthogonal matrix is defined as a square matrix having its inverse equal to its transpose, so that if

$$\mathbf{A}^{-1} = \mathbf{A}^{\mathsf{T}} \tag{4-22}$$

then **A** is an orthogonal matrix.

An example of an orthogonal matrix of second order is the following:

$$\mathbf{A} = \begin{bmatrix} 0.6 & -0.8 \\ 0.8 & 0.6 \end{bmatrix}$$

The transpose of the matrix **A** is

$$\mathbf{A}^{\mathsf{T}} = \begin{bmatrix} 0.6 & 0.8 \\ -0.8 & 0.6 \end{bmatrix}$$

and when **A** and $\mathbf{A}^{\mathsf{T}}$ are multiplied (in either order) the identity matrix is obtained. Therefore, the transpose of the matrix is indeed its inverse.

Examination of the matrix **A** in the above example shows that each row of the matrix is a unit vector, since

$$(0.6)^2 + (0.8)^2 = 1$$

Furthermore, the scalar product of the first row and the second row (each considered as a vector) is zero:

$$0.6(0.8) - 0.8(0.6) = 0$$

Thus, the rows of the matrix are orthogonal unit vectors. The same two conditions hold for the transposed matrix, which means that the columns of the original matrix are also orthogonal unit vectors.

A set of orthogonal unit vectors, such as the row and column vectors in the above example, is said to be *orthonormal*. Thus, for an orthonormal set of vectors the following relationships hold in general:

$$\begin{aligned} \mathbf{V}_i\mathbf{V}_j^{\mathsf{T}} &= 0 \quad \text{if} \quad i \neq j \\ \mathbf{V}_i\mathbf{V}_j^{\mathsf{T}} &= 1 \quad \text{if} \quad i = j \end{aligned} \tag{4-23}$$

in which $\mathbf{V}_i$ and $\mathbf{V}_j$ are n-dimensional row vectors. The above pair of equations can be written more concisely by introducing the *Kronecker delta* δ_{ij}, defined as*

$$\begin{aligned} \delta_{ij} &= 0 \quad \text{if} \quad i \neq j \\ \delta_{ij} &= 1 \quad \text{if} \quad i = j \end{aligned} \tag{4-24}$$

*Leopold Kronecker (1823–1891) was a German mathematician.

With this notation, the conditions for an orthonormal set of vectors become

$$\mathbf{V}_i\mathbf{V}_j^{\mathsf{T}} = \delta_{ij} \tag{4-25}$$

If the rows of a matrix constitute an orthonormal set of vectors, then the columns automatically will be an orthonormal set, and the matrix itself will be orthogonal, as illustrated in the 2 × 2 example given previously. To show this hypothesis in general, consider an nth order square matrix **A** in which the row vectors **V** satisfy Equation (4-25):

$$\mathbf{A} = \begin{bmatrix} \mathbf{V}_1 \\ \mathbf{V}_2 \\ \cdots \\ \mathbf{V}_n \end{bmatrix}$$

In this matrix, each row vector has dimension n. Now suppose that **A** is post-multiplied by its transpose, which is a matrix that has the transposes of the vectors $\mathbf{V}_1, \mathbf{V}_2, \ldots, \mathbf{V}_n$ as its columns:

$$\mathbf{A}^{\mathsf{T}} = [\mathbf{V}_1^{\mathsf{T}} \quad \mathbf{V}_2^{\mathsf{T}} \quad \ldots \quad \mathbf{V}_n^{\mathsf{T}}]$$

When the product $\mathbf{AA}^{\mathsf{T}}$ is taken, the identity matrix is obtained:

$$\mathbf{AA}^{\mathsf{T}} = \begin{bmatrix} \mathbf{V}_1\mathbf{V}_1^{\mathsf{T}} & \mathbf{V}_1\mathbf{V}_2^{\mathsf{T}} & \ldots & \mathbf{V}_1\mathbf{V}_n^{\mathsf{T}} \\ \mathbf{V}_2\mathbf{V}_1^{\mathsf{T}} & \mathbf{V}_2\mathbf{V}_2^{\mathsf{T}} & \ldots & \mathbf{V}_2\mathbf{V}_n^{\mathsf{T}} \\ \ldots & \ldots & \ldots & \ldots \\ \mathbf{V}_n\mathbf{V}_1^{\mathsf{T}} & \mathbf{V}_n\mathbf{V}_2^{\mathsf{T}} & \ldots & \mathbf{V}_n\mathbf{V}_n^{\mathsf{T}} \end{bmatrix} = \mathbf{I}$$

The identity matrix is also obtained from the product $\mathbf{A}^{\mathsf{T}}\mathbf{A}$, because the colums of **A** are an orthonormal set of vectors. Therefore, the inverse of **A** is its transpose, and **A** is an orthogonal matrix. The fact that the rows of the matrix are orthogonal unit vectors is both a necessary and sufficient condition for the matrix to be orthogonal.

Example. Verify the fact that the matrix **A** shown below is an orthogonal matrix:

$$\mathbf{A} = \begin{bmatrix} 0.283 & -0.572 & -0.770 \\ -0.572 & 0.544 & -0.614 \\ 0.770 & 0.614 & -0.173 \end{bmatrix}$$

The product of **A** and its transpose is

$$\mathbf{AA^T} = \begin{bmatrix} 0.283 & -0.572 & -0.770 \\ -0.572 & 0.544 & -0.614 \\ 0.770 & 0.614 & -0.173 \end{bmatrix} \begin{bmatrix} 0.283 & -0.572 & 0.770 \\ -0.572 & 0.544 & 0.614 \\ -0.770 & -0.614 & -0.173 \end{bmatrix}$$

$$= \begin{bmatrix} 1.000 & 0 & 0 \\ 0 & 1.000 & 0 \\ 0 & 0 & 1.000 \end{bmatrix} = \mathbf{I}$$

This result shows that the transpose of **A** is its inverse; therefore, **A** is an orthogonal matrix.

Because **A** is orthogonal, its rows must constitute a set of orthonormal vectors satisfying Equation (4–25). For example, the inner product of the first row with itself is

$$(0.283)^2 + (-0.572)^2 + (-0.770)^2 = 1.000 \tag{a}$$

and the inner product of the first and second rows is

$$0.283(-0.572) - 0.572(0.544) - 0.770(-0.614) = 0 \tag{b}$$

Similar results are obtained for the other combinations of rows and also for the columns of the matrix. Of course, the calculations performed in Equations (a) and (b) constitute two of the inner products involved in calculating the product $\mathbf{AA^T}$ above.

If two square matrices $\mathbf{A}_1$ and $\mathbf{A}_2$ are orthogonal, then the product $\mathbf{A}_1\mathbf{A}_2$ will also be an orthogonal matrix. Application of Equations (4–17), (4–22), and (2–19a) in sequence shows the validity of this statement as follows:

$$(\mathbf{A}_1\mathbf{A}_2)^{-1} = \mathbf{A}_2^{-1}\mathbf{A}_1^{-1} = \mathbf{A}_2^{\mathsf{T}}\mathbf{A}_1^{\mathsf{T}} = (\mathbf{A}_1\mathbf{A}_2)^{\mathsf{T}} \tag{4–26}$$

Indeed, the product of any number of orthogonal matrices will result in an orthogonal product matrix. Hence, the following relationship holds:

$$(\mathbf{A}_1\mathbf{A}_2 \ldots \mathbf{A}_n)^{-1} = (\mathbf{A}_1\mathbf{A}_2 \ldots \mathbf{A}_n)^{\mathsf{T}} \tag{4–27}$$

in which $\mathbf{A}_1, \mathbf{A}_2, \ldots \mathbf{A}_n$ are orthogonal matrices.

It is also interesting to observe that a symmetric orthogonal matrix is its own inverse, because a symmetric matrix is equal to its transpose. An example is the following matrix:

$$\mathbf{A} = \begin{bmatrix} \sin\alpha & \cos\alpha \\ \cos\alpha & -\sin\alpha \end{bmatrix}$$

for which it is easy to demonstrate that $\mathbf{A}^{-1} = \mathbf{A}^{\mathsf{T}} = \mathbf{A}$. Other examples are given in Problems 4.6-5 and 4.6-8.

The determinant of an orthogonal matrix must be equal to the square root of unity (that is, either plus one or minus one). The reason is that since $\mathbf{A}\mathbf{A}^{\mathsf{T}} = \mathbf{I}$, the product of the determinant of $\mathbf{A}$ and the determinant of $\mathbf{A}^{\mathsf{T}}$ must be unity (see Equation 4-21). However, the value of the determinant of a matrix is not changed by transposition (see Equation 3-5), therefore, the values of both $|\mathbf{A}|$ and $|\mathbf{A}^{\mathsf{T}}|$ must be the same and equal to $+1$ or -1.

4.7 Inversion by Successive Transformations. One of the procedures for calculating the inverse of a matrix is the method of *successive transformations*. This method involves the construction of transformation matrices (see Section 2.6) that transform, by means of successive operations, the given matrix into the identity matrix. Once the transformation matrices have been properly obtained, the inverse of the matrix will be equal to their product. For example, suppose that the given matrix is $\mathbf{A}$ and that $\mathbf{T}_1, \mathbf{T}_2, \ldots, \mathbf{T}_n$ are transformation matrices such that when $\mathbf{A}$ is premultiplied by them in the order shown below, the result is the identity matrix:

$$\mathbf{T}_n \ldots \mathbf{T}_2 \mathbf{T}_1 \mathbf{A} = \mathbf{I} \tag{4-28}$$

From this equation it follows that the inverse of $\mathbf{A}$ is the product

$$\mathbf{A}^{-1} = \mathbf{T}_n \ldots \mathbf{T}_2 \mathbf{T}_1 \tag{4-29}$$

If the transformation matrices are premultipliers, as in Equation (4-28), the successive operations are made on the rows of $\mathbf{A}$. However, it is equally feasible to perform the operations on the columns of $\mathbf{A}$ by postmultiplication with appropriate transformation matrices.

The essential feature of the method of successive transformations is the determination of the transformation matrices themselves. In order to show how they are obtained, assume that the following matrix is to be inverted:

$$\mathbf{A} = \begin{bmatrix} A_{11} & A_{12} & \ldots & A_{1n} \\ A_{21} & A_{22} & \ldots & A_{2n} \\ \ldots & \ldots & \ldots & \ldots \\ A_{n1} & A_{n2} & \ldots & A_{nn} \end{bmatrix} \tag{4-30}$$

Also, assume that the transformation matrices are to be premultipliers that perform operations on the rows of $\mathbf{A}$.

The first transformation matrix is constructed so that it will transform the first column of $\mathbf{A}$ into the first column of the identity matrix; this result can be accomplished by dividing the first row of $\mathbf{A}$ by A_{11} and then adding appropriate multiples of the first row to the other rows. Thus, the elementary operations required are scaling and combining (see Section 2.6). The transformation matrix

$\mathbf{T}_1$, which performs both the scaling and the combining, is

$$\mathbf{T}_1 = \begin{bmatrix} \dfrac{1}{A_{11}} & 0 & \dots & 0 \\ -\dfrac{A_{21}}{A_{11}} & 1 & \dots & 0 \\ \dots & \dots & \dots & \dots \\ -\dfrac{A_{n1}}{A_{11}} & 0 & \dots & 1 \end{bmatrix} \tag{4-31}$$

Note that $\mathbf{T}_1$ is found by beginning with the identity matrix and then performing on it the operations that are desired for the matrix $\mathbf{A}$. Specifically, the first row in the identity matrix is divided by A_{11} in order to produce a value of unity in the leading position of the matrix $\mathbf{A}$. The element selected for transformation into unity (in this case, the element A_{11}) is called the *pivot element*. Multiples of the pivot element must be added to the remaining elements in the first column in order to produce zeros. For instance, minus A_{21} times the pivot element (which now has the value unity in the matrix $\mathbf{A}$) must be added to the first element in the second row. To accomplish this result, minus A_{21} times the first row of $\mathbf{T}_1$ is added to the second row, as shown in Equation (4-31). The other elements of the first column of $\mathbf{T}_1$ are found in the same fashion.

If the matrix $\mathbf{A}$ is premultiplied by $\mathbf{T}_1$ the result is

$$\mathbf{T}_1\mathbf{A} = \begin{bmatrix} 1 & \dfrac{A_{12}}{A_{11}} & \dots & \dfrac{A_{1n}}{A_{11}} \\ 0 & A_{22} - \dfrac{A_{21}}{A_{11}} A_{12} & \dots & A_{2n} - \dfrac{A_{21}}{A_{11}} A_{1n} \\ \dots & \dots & \dots & \dots \\ 0 & A_{n2} - \dfrac{A_{n1}}{A_{11}} A_{12} & \dots & A_{nn} - \dfrac{A_{n1}}{A_{11}} A_{1n} \end{bmatrix} \tag{4-32a}$$

or, using new symbols for the transformed matrix,

$$\mathbf{T}_1\mathbf{A} = \begin{bmatrix} 1 & B_{12} & \dots & B_{1n} \\ 0 & B_{22} & \dots & B_{2n} \\ \dots & \dots & \dots & \dots \\ 0 & B_{n2} & \dots & B_{nn} \end{bmatrix} = \mathbf{B} \tag{4-32b}$$

Next, a transformation matrix $\mathbf{T}_2$ that changes the second column of the matrix **B** into the second column of the identity matrix is formed, using element B_{22} as the pivot:

$$\mathbf{T}_2 = \begin{bmatrix} 1 & -\dfrac{B_{12}}{B_{22}} & \dots & 0 \\ 0 & \dfrac{1}{B_{22}} & \dots & 0 \\ \dots & \dots & \dots & \dots \\ 0 & -\dfrac{B_{n2}}{B_{22}} & \dots & 1 \end{bmatrix} \tag{4-33}$$

Premultiplication of **B** by $\mathbf{T}_2$ gives a new matrix **C**, in which the first two columns correspond to the identity matrix:

$$\mathbf{T}_2\mathbf{B} = \mathbf{T}_2\mathbf{T}_1\mathbf{A} = \begin{bmatrix} 1 & 0 & \dots & C_{1n} \\ 0 & 1 & \dots & C_{2n} \\ \dots & \dots & \dots & \dots \\ 0 & 0 & \dots & C_{nn} \end{bmatrix} = \mathbf{C} \tag{4-34}$$

The process of transforming columns is applied a total of n times in the manner illustrated until the product becomes the identity matrix:

$$\mathbf{T}_n \dots \mathbf{T}_2\mathbf{T}_1\mathbf{A} = \mathbf{I} \tag{4-28 repeated}$$

Thus, the inverse of **A** is the product of all of the transformation matrices (see Equation 4-29).

Some examples will now be given to illustrate the method of successive transformations in the form described above. Later, a more efficient method for recording the calculations is demonstrated. The latter method eliminates the necessity of actually performing any matrix multiplications and is illustrated in Examples 3, 4, and 5.

Example 1. The following second-order matrix is to be inverted:

$$\mathbf{A} = \begin{bmatrix} 2 & -1 \\ 3 & -4 \end{bmatrix}$$

This matrix is used as an example solely to illustrate the successive transformation method; in practice, it is easier to invert a second-order matrix by using Equation (4-12). However, more realistic examples involving third- and fourth-order matrices are given later.

It is assumed first that the inversion of the matrix **A** is to be accomplished by operating on the rows; therefore, the transformation matrices will be premultipliers of **A**. As explained above, the matrix $\mathbf{T}_1$ is constructed so that it will transform the first element of the first row of **A** into unity and will transform the first element of the second row into zero. Referring to Equation (4-31) for this purpose, the matrix $\mathbf{T}_1$ becomes:

$$\mathbf{T}_1 = \begin{bmatrix} \dfrac{1}{A_{11}} & 0 \\ -\dfrac{A_{21}}{A_{11}} & 1 \end{bmatrix} = \begin{bmatrix} 0.5 & 0 \\ -1.5 & 1 \end{bmatrix}$$

and the product $\mathbf{T}_1\mathbf{A}$, denoted as **B**, is

$$\mathbf{T}_1\mathbf{A} = \begin{bmatrix} 1 & -0.5 \\ 0 & -2.5 \end{bmatrix} = \begin{bmatrix} 1 & B_{12} \\ 0 & B_{22} \end{bmatrix} = \mathbf{B}$$

It is apparent that the first transformation has produced the desired results in the first column.

Transformation of the matrix **B** into the identity matrix requires the use of a second transformation matrix $\mathbf{T}_2$ of the type given by Equation (4-33):

$$\mathbf{T}_2 = \begin{bmatrix} 1 & -\dfrac{B_{12}}{B_{22}} \\ 0 & \dfrac{1}{B_{22}} \end{bmatrix} = \begin{bmatrix} 1 & -0.2 \\ 0 & -0.4 \end{bmatrix}$$

Then the product $\mathbf{T}_2\mathbf{B}$ becomes the identity matrix:

$$\mathbf{T}_2\mathbf{B} = \mathbf{T}_2\mathbf{T}_1\mathbf{A} = \begin{bmatrix} 1 & 0 \\ 0 & 1 \end{bmatrix} = \mathbf{I}$$

Thus, the desired transformation of the matrix **A** into the identity matrix has been carried out, and the inverse of **A** is

$$\mathbf{A}^{-1} = \mathbf{T}_2\mathbf{T}_1 = \begin{bmatrix} 1 & -0.2 \\ 0 & -0.4 \end{bmatrix} \begin{bmatrix} 0.5 & 0 \\ -1.5 & 1 \end{bmatrix} = \begin{bmatrix} 0.8 & -0.2 \\ 0.6 & -0.4 \end{bmatrix}$$

When transforming by premultiplication as in the calculations above, the transformation matrices perform operations on the rows of the matrix. However, the first matrix produces the desired results (the first value equal to unity; the remaining values equal to zero) in the first column of the matrix; the second matrix produces the desired results in the second column; and so on until all columns of the matrix have been transformed into those of the identity matrix.

A similar procedure can be carried out by postmultiplication of **A** with suitable transformation matrices. In this case, the operations are performed on the columns of the original matrix. However, the function of the first transformation matrix is to produce the desired results (the first value equal to unity, the remaining values equal to zero) in the first row of the matrix; the second matrix produces the desired results in the second row; and so forth. To illustrate the procedure, the second-order matrix **A** in this example will be inverted again using column operations. The transformation matrix $\mathbf{T}_3$ required to transform the first row of **A** to the identity matrix is

$$\mathbf{T}_3 = \begin{bmatrix} \dfrac{1}{A_{11}} & -\dfrac{A_{12}}{A_{11}} \\ 0 & 1 \end{bmatrix} = \begin{bmatrix} 0.5 & 0.5 \\ 0 & 1 \end{bmatrix}$$

and the product $\mathbf{AT}_3$, denoted as **D**, is

$$\mathbf{AT}_3 = \begin{bmatrix} 1 & 0 \\ 1.5 & -2.5 \end{bmatrix} = \begin{bmatrix} 1 & 0 \\ D_{21} & D_{22} \end{bmatrix} = \mathbf{D}$$

Thus, the desired transformation of the first row is obtained. Next, the transformation matrix $\mathbf{T}_4$ is constructed for the purpose of transforming the second row of **D** to the identity matrix:

$$\mathbf{T}_4 = \begin{bmatrix} 1 & 0 \\ -\dfrac{D_{21}}{D_{22}} & \dfrac{1}{D_{22}} \end{bmatrix} = \begin{bmatrix} 1 & 0 \\ 0.6 & -0.4 \end{bmatrix}$$

Postmultiplication of **D** by $\mathbf{T}_4$ produces the identity matrix as follows:

$$\mathbf{DT}_4 = \mathbf{AT}_3\mathbf{T}_4 = \begin{bmatrix} 1 & 0 \\ 0 & 1 \end{bmatrix} = \mathbf{I}$$

and the inverse of the matrix **A** is the product $\mathbf{T}_3\mathbf{T}_4$:

$$\mathbf{A}^{-1} = \mathbf{T}_3\mathbf{T}_4 = \begin{bmatrix} 0.5 & 0.5 \\ 0 & 1 \end{bmatrix} \begin{bmatrix} 1 & 0 \\ 0.6 & -0.4 \end{bmatrix} = \begin{bmatrix} 0.8 & -0.2 \\ 0.6 & -0.4 \end{bmatrix}$$

which is the same result found previously.

Example 2. As a second example of the method of successive transformations, assume that the following 3 × 3 matrix **A** is to be inverted using operations on rows:

$$\mathbf{A} = \begin{bmatrix} 20 & -14 & 3 \\ -2 & 1 & 0 \\ 5 & -4 & 1 \end{bmatrix}$$

The first transformation matrix $\mathbf{T}_1$ is determined from Equation (4-31) as follows:

$$\mathbf{T}_1 = \begin{bmatrix} 0.05 & 0 & 0 \\ 0.10 & 1 & 0 \\ -0.25 & 0 & 1 \end{bmatrix}$$

and the product $\mathbf{T}_1\mathbf{A}$ is

$$\mathbf{B} = \mathbf{T}_1\mathbf{A} = \begin{bmatrix} 1 & -0.7 & 0.15 \\ 0 & -0.4 & 0.30 \\ 0 & -0.5 & 0.25 \end{bmatrix}$$

The second transformation matrix $\mathbf{T}_2$ follows the form given by Equation (4-33); thus,

$$\mathbf{T}_2 = \begin{bmatrix} 1 & -1.75 & 0 \\ 0 & -2.50 & 0 \\ 0 & -1.25 & 1 \end{bmatrix}$$

The next product in the solution is

$$\mathbf{C} = \mathbf{T}_2\mathbf{T}_1\mathbf{A} = \begin{bmatrix} 1 & 0 & -0.375 \\ 0 & 1 & -0.750 \\ 0 & 0 & -0.125 \end{bmatrix}$$

The third transformation matrix can be constructed in a similar manner and is as follows:

$$\mathbf{T}_3 = \begin{bmatrix} 1 & 0 & -3 \\ 0 & 1 & -6 \\ 0 & 0 & -8 \end{bmatrix}$$

The final product $\mathbf{T}_3\mathbf{T}_2\mathbf{T}_1\mathbf{A}$ produces the identity matrix, as can be confirmed easily. Hence, the inverse of the matrix $\mathbf{A}$ is determined by multiplying the three transformation matrices as follows:

$$\mathbf{A}^{-1} = \mathbf{T}_3\mathbf{T}_2\mathbf{T}_1 = \begin{bmatrix} 1 & 2 & -3 \\ 2 & 5 & -6 \\ 3 & 10 & -8 \end{bmatrix}$$

The matrix $\mathbf{A}^{-1}$ found above can be obtained equally well by operating on the columns of $\mathbf{A}$. It is left as an exercise for the reader to perform the inversion in that manner.

It is always possible to write out in detail the various transformation matrices, as illustrated in the two preceding examples, and then to obtain the matrix inverse from their product. However, the calculations are somewhat tedious when performed in that manner, and it is desirable to organize them into a more systematic calculating scheme. When the operations are performed on the rows, the procedure is to place the identity matrix to the right of the matrix to be inverted, and then to perform on it the same transformations that are performed on the given matrix. When the transformations have been completed, the identity matrix will appear on the left and the inverse matrix on the right. If the operations are performed on the columns, the identity matrix is placed below the given matrix. Then, after the transformations have been performed, the identity matrix will be above and the inverse matrix below. The following examples illustrate this method of calculation.

Example 3. The second-order matrix **A** in Example 1 will be used again. In the first inversion, the operations are on the rows, and in the second the operations are on the columns. For the former case, the first step is to write the identity matrix to the right of the matrix to be inverted:

$$\begin{bmatrix} 2 & -1 \\ 3 & -4 \end{bmatrix} \quad \begin{bmatrix} 1 & 0 \\ 0 & 1 \end{bmatrix}$$

The same transformations as those in Example 1 will now be made without actually forming the transformation matrices or executing the matrix multiplications. Each transformation will be performed not only on the rows of **A**, but also on the rows of the identity matrix. In order to transform the first element of the first row of **A** into unity, the first row must be divided by 2. Then, transforming the first element of the second row into zero requires that three times the first row be subtracted from the second row. When these two steps are performed on both matrices, the results are

$$\begin{bmatrix} 1 & -0.5 \\ 0 & -2.5 \end{bmatrix} \quad \begin{bmatrix} 0.5 & 0 \\ -1.5 & 1 \end{bmatrix}$$

Next, the second row must be divided by -2.5, and 0.5 times the second row is then added to the first row. The result is

$$\begin{bmatrix} 1 & 0 \\ 0 & 1 \end{bmatrix} \quad \begin{bmatrix} 0.8 & -0.2 \\ 0.6 & -0.4 \end{bmatrix}$$

The final result of these operations is that the original matrix **A** (on the left-hand side) has been transformed into the identity matrix, while the identity matrix (on the right-hand side) has been transformed into the inverse of **A**. The reader can compare these calculations with those of the earlier example in order to note that the same calculations are being performed in both cases, but the new solution is more efficient.

When the inversion is to be carried out by operating on the columns of the matrix instead of the rows, the identity matrix is placed below the matrix to be inverted:

$$\begin{bmatrix} 2 & -1 \\ 3 & -4 \end{bmatrix}$$

$$\begin{bmatrix} 1 & 0 \\ 0 & 1 \end{bmatrix}$$

As the first step in the transformation, the first column is divided by 2, and then the first column is added to the second column:

$$\begin{bmatrix} 1 & 0 \\ 1.5 & -2.5 \end{bmatrix}$$

$$\begin{bmatrix} 0.5 & 0.5 \\ 0 & 1 \end{bmatrix}$$

Next, the second column is divided by -2.5, and -1.5 times the second column is added to the first column:

$$\begin{bmatrix} 1 & 0 \\ 0 & 1 \end{bmatrix}$$

$$\begin{bmatrix} 0.8 & -0.2 \\ 0.6 & -0.4 \end{bmatrix}$$

Again it is seen that the operations which convert the original matrix into the identity matrix will also convert the identity matrix into the inverse of the original matrix. When performing the calculations in tabular form, the brackets denoting the matrices may be omitted.

Example 4. The matrix of order three shown below on the left-hand side is to be inverted by successive transformations using row operations. The identity matrix has been placed to the right of the given matrix:

$$\begin{bmatrix} -2 & 3 & 1 \\ -5 & 4 & 2 \\ 3 & 0 & -1 \end{bmatrix} \begin{bmatrix} 1 & 0 & 0 \\ 0 & 1 & 0 \\ 0 & 0 & 1 \end{bmatrix}$$

In order to transform the first column of **A** to the identity matrix, the following series of three operations must be executed: The first row is divided by minus two; five times the first row is added to the second row; and then three times the first row must be subtracted from the third row. The result of performing these operations on both matrices is

$$\begin{bmatrix} 1 & -3/2 & -1/2 \\ 0 & -7/2 & -1/2 \\ 0 & 9/2 & 1/2 \end{bmatrix} \begin{bmatrix} -1/2 & 0 & 0 \\ -5/2 & 1 & 0 \\ 3/2 & 0 & 1 \end{bmatrix}$$

Next, a similar series of three operations is performed with the second row. In this case, the second element of the second row is the pivot element that must be transformed to unity, while the other elements of the second column are transformed to zero. These results are accomplished by dividing the second row by $-7/2$, adding $3/2$ times the second row to the first row, and subtracting $9/2$ times the second row from the third row. Performing these three operations on both matrices given above yields the following:

$$\begin{bmatrix} 1 & 0 & -2/7 \\ 0 & 1 & 1/7 \\ 0 & 0 & -1/7 \end{bmatrix} \begin{bmatrix} 4/7 & -3/7 & 0 \\ 5/7 & -2/7 & 0 \\ -12/7 & 9/7 & 1 \end{bmatrix}$$

The last transformation consists of dividing the third row by $-1/7$, adding $2/7$ times the third row to the first row, and subtracting $1/7$ times the third row from the second row:

$$\begin{bmatrix} 1 & 0 & 0 \\ 0 & 1 & 0 \\ 0 & 0 & 1 \end{bmatrix} \begin{bmatrix} 4 & -3 & -2 \\ -1 & 1 & 1 \\ 12 & -9 & -7 \end{bmatrix}$$

The successive transformations have now been completed; the identity matrix appears on the left, and the inverse of the original matrix is on the right. The same result for the inverse can be obtained by operations on columns instead of rows.

Example 5. The matrix of order four shown below on the left-hand side is to be inverted by successive transformations using row operations.

$$\begin{bmatrix} 2.847 & 1.721 & 1.638 & 0.843 \\ 2.012 & 0.945 & 4.237 & 1.658 \\ 1.013 & 1.426 & 2.166 & 0.881 \\ 0.980 & 0.879 & 2.257 & 1.743 \end{bmatrix} \begin{bmatrix} 1.000 & 0 & 0 & 0 \\ 0 & 1.000 & 0 & 0 \\ 0 & 0 & 1.000 & 0 \\ 0 & 0 & 0 & 1.000 \end{bmatrix}$$

The first series of transformations is as follows. The first row will be divided by 2.847 in order to transform the first element into unity; the first row times -2.012 will then be added to the second row in order to transform the first element of the second row to zero; the first row times -1.013 will be added to the third row for the same reason; and, lastly, −0.980 times the first row will be added to the fourth row. The result of these four row transformations is

$$\begin{bmatrix} 1.000 & 0.604 & 0.575 & 0.296 \\ 0 & -0.271 & 3.079 & 1.062 \\ 0 & 0.814 & 1.583 & 0.581 \\ 0 & 0.287 & 1.693 & 1.453 \end{bmatrix} \begin{bmatrix} 0.351 & 0 & 0 & 0 \\ -0.707 & 1.000 & 0 & 0 \\ -0.356 & 0 & 1.000 & 0 \\ -0.344 & 0 & 0 & 1.000 \end{bmatrix}$$

Next, a similar series of transformations is performed with the second row. In this case the second row is divided by −0.271, and then appropriate ratios times the second row are added to rows one, three, and four. The result of these operations is the following:

$$\begin{bmatrix} 1.000 & 0 & 7.438 & 2.663 \\ 0 & 1.000 & -11.353 & -3.916 \\ 0 & 0 & 10.820 & 3.767 \\ 0 & 0 & 4.947 & 2.575 \end{bmatrix} \begin{bmatrix} -1.224 & 2.229 & 0 & 0 \\ 2.605 & -3.687 & 0 & 0 \\ -2.476 & 3.000 & 1.000 & 0 \\ -1.091 & 1.057 & 0 & 1.000 \end{bmatrix}$$

When a similar series of operations is performed with the third row, the result is

$$\begin{bmatrix} 1.000 & 0 & 0 & 0.074 \\ 0 & 1.000 & 0 & 0.037 \\ 0 & 0 & 1.000 & 0.348 \\ 0 & 0 & 0 & 0.853 \end{bmatrix} \begin{bmatrix} 0.478 & 0.167 & -0.687 & 0 \\ 0.008 & -0.539 & 1.049 & 0 \\ -0.229 & 0.277 & 0.092 & 0 \\ 0.041 & -0.315 & -0.457 & 1.000 \end{bmatrix}$$

Finally, the results obtained from operating with the fourth row are

$$\begin{bmatrix} 1.000 & 0 & 0 & 0 \\ 0 & 1.000 & 0 & 0 \\ 0 & 0 & 1.000 & 0 \\ 0 & 0 & 0 & 1.000 \end{bmatrix} \begin{bmatrix} 0.475 & 0.194 & -0.648 & -0.086 \\ 0.006 & -0.526 & 1.069 & -0.043 \\ -0.246 & 0.406 & 0.279 & -0.408 \\ 0.048 & -0.369 & -0.536 & 1.173 \end{bmatrix}$$

Thus, the inverse of the matrix has been obtained, and its correctness can be checked by multiplying it by the original matrix. The result of the multiplication will be a matrix that is nearly the identity matrix, depending upon the accuracy of the calculations.

In all of the preceding examples, the first pivot was selected as the leading element, and then successive pivots were taken in sequence along the principal diagonal. This sequence is arbitrary, and it is permissible to select the pivots in some other order if desired.

If a pivot element has the value zero, then it is necessary to interchange either two rows or two columns in order to produce a nonzero pivot element. Interchanging two rows of the original matrix will cause the same interchange to take place in the columns of the inverse, and an interchange of two columns of the original matrix results in a corresponding interchange of rows of the inverse (see Section 4.5). Hence, after the inversion is completed, it is necessary to make appropriate interchanges in the calculated inverse in order to obtain the inverse of the given matrix. For example, suppose that a matrix **A** is to be inverted and that prior to inversion the second and fourth rows are interchanged. After inversion of the altered matrix, the second and fourth columns of the calculated inverse must be interchanged to give the true inverse of **A**.

The numerical accuracy obtained by the successive transformation method can be improved by selecting the numerically largest elements in the matrix as pivot elements. This procedure requires interchanging rows and columns of the matrix prior to each successive transformation until the largest element is in pivot position. Since interchanges may occur prior to the selection of each pivot element, it is necessary to keep account of all such interchanges during the process of inversion. After the inversion is completed, the calculated inverse must be rearranged to give the true inverse.

It was tacitly assumed in the examples of this section that the original matrix was not singular. Singularity can always be checked by calculating the determinant of the matrix before beginning the inversion. If this is not done, however, the existence of singularity will be discovered during the process of calculating the inverse. It will be found that it is impossible to perform the successive transformation calculations, because at some stage a difficulty (such as division by zero) will be encountered.

4.8. Inversion by Partitioning. In the method described in this section, the inversion of a given matrix is replaced by the inversion, multiplication, and subtraction of several smaller matrices. The method is useful in the case of hand calculations when the original matrix is too large for direct inversion to be feasible. It is also useful for computer calculations when the matrix is so large that it cannot be stored properly in the computer memory and must be broken down into smaller parts.

Suppose that a square matrix **P** of order n is to be inverted. This matrix can be partitioned into four submatrices **A**, **B**, **C**, and **D** such that the diagonal submatrices **A** and **D** are square; thus

$$\mathbf{P} = \begin{bmatrix} \mathbf{A} & \mathbf{B} \\ \mathbf{C} & \mathbf{D} \end{bmatrix} \tag{4-35}$$

It is assumed that **A** is a square matrix of order p, **D** is a square matrix of order q, **C** is of size $q \times p$, and **B** is of size $p \times q$, where $p + q = n$.

The inverse matrix $\mathbf{P}^{-1}$ will be partitioned in exactly the same manner as the original matrix:

$$\mathbf{P}^{-1} = \begin{bmatrix} \mathbf{Q} & \mathbf{R} \\ \mathbf{S} & \mathbf{T} \end{bmatrix} \tag{4-36}$$

so that $\mathbf{Q}$ is $p \times p$, $\mathbf{T}$ is $q \times q$, $\mathbf{S}$ is $q \times p$, and $\mathbf{R}$ is $p \times q$.

From the definition of an inverse (see Equation 4–1) and the rules for multiplying partitioned matrices (see Section 2.7), the following four equations are obtained:

$$\mathbf{AQ} + \mathbf{BS} = \mathbf{I} \tag{a}$$

$$\mathbf{AR} + \mathbf{BT} = \mathbf{0} \tag{b}$$

$$\mathbf{CQ} + \mathbf{DS} = \mathbf{0} \tag{c}$$

$$\mathbf{CR} + \mathbf{DT} = \mathbf{I} \tag{d}$$

These equations can be solved for the submatrices $\mathbf{Q}$, $\mathbf{R}$, $\mathbf{S}$, and $\mathbf{T}$ in the following way. Premultiply Equation (c) by $\mathbf{BD}^{-1}$ and subtract the result from Equation (a), yielding

$$(\mathbf{A} - \mathbf{BD}^{-1}\mathbf{C})\mathbf{Q} = \mathbf{I} \qquad \text{or} \qquad \mathbf{Q} = (\mathbf{A} - \mathbf{BD}^{-1}\mathbf{C})^{-1}$$

Next, Equations (c) and (d) can be solved for submatrices $\mathbf{S}$ and $\mathbf{T}$, respectively:

$$\mathbf{S} = -\mathbf{D}^{-1}\mathbf{CQ} \qquad \mathbf{T} = \mathbf{D}^{-1} - \mathbf{D}^{-1}\mathbf{CR}$$

Finally, submatrix $\mathbf{R}$ is found by substituting the preceding expression for submatrix $\mathbf{T}$ into Equation (b) and solving:

$$\mathbf{R} = -(\mathbf{A} - \mathbf{BD}^{-1}\mathbf{C})^{-1}\mathbf{BD}^{-1} = -\mathbf{QBD}^{-1}$$

Thus, the submatrices $\mathbf{Q}$, $\mathbf{R}$, $\mathbf{S}$, and $\mathbf{T}$ in the matrix inverse are obtained by solving sequentially the following four equations:

$$\mathbf{Q} = (\mathbf{A} - \mathbf{BD}^{-1}\mathbf{C})^{-1} \tag{4-37a}$$

$$\mathbf{R} = -\mathbf{QBD}^{-1} \tag{4-37b}$$

$$\mathbf{S} = -\mathbf{D}^{-1}\mathbf{CQ} \tag{4-37c}$$

$$\mathbf{T} = \mathbf{D}^{-1} - \mathbf{D}^{-1}\mathbf{CR} \tag{4-37d}$$

The use of these equations requires that the submatrix $\mathbf{D}$ and the term in parentheses in the first equation be nonsingular. Note that two inversions are required, one of order $p \times p$ and the other of order $q \times q$.

An alternate approach, requiring the inversion of submatrix $\mathbf{A}$ instead of $\mathbf{D}$, is obtained by solving Equations (a), (b), (c), and (d) in a different sequence.

The first step is to premultiply Equation (b) by $\mathbf{CA}^{-1}$ and subtract from Equation (d) to obtain

$$\mathbf{T} = (\mathbf{D} - \mathbf{CA}^{-1}\mathbf{B})^{-1}$$

Then Equations (b) and (a) yield

$$\mathbf{R} = -\mathbf{A}^{-1}\mathbf{BT} \qquad \mathbf{Q} = \mathbf{A}^{-1} - \mathbf{A}^{-1}\mathbf{BS}$$

Finally, substitute submatrix $\mathbf{Q}$ into Equation (c) and solve:

$$\mathbf{S} = -\mathbf{TCA}^{-1}$$

Again, four equations that can be solved sequentially are obtained:

$$\mathbf{T} = (\mathbf{D} - \mathbf{CA}^{-1}\mathbf{B})^{-1} \tag{4-38a}$$

$$\mathbf{S} = -\mathbf{TCA}^{-1} \tag{4-38b}$$

$$\mathbf{R} = -\mathbf{A}^{-1}\mathbf{BT} \tag{4-38c}$$

$$\mathbf{Q} = \mathbf{A}^{-1} - \mathbf{A}^{-1}\mathbf{BS} \tag{4-38d}$$

As before, two inversions are required, one of order $p \times p$ and the other of order $q \times q$. Equations (4-37) and (4-38) can be verified by multiplying $\mathbf{P}$ by $\mathbf{P}^{-1}$ to obtain the identity matrix.

If both submatrices $\mathbf{A}$ and $\mathbf{D}$ are nonsingular, then either set of equations (Equations 4-37 or 4-38) may be used to obtain the inverse; there is no advantage to be gained by selecting one set of equations in preference to the other. Of course, if either $\mathbf{A}$ or $\mathbf{D}$ is singular, only one set of equations will be applicable. If both $\mathbf{A}$ and $\mathbf{D}$ are singular, neither set of formulas can be used; the remedies are to change the partitioning until either $\mathbf{A}$ or $\mathbf{D}$ becomes nonsingular or to use another method for obtaining the inverse.

Important special cases arise when one or both of the off-diagonal submatrices $\mathbf{B}$ and $\mathbf{C}$ are null matrices. If $\mathbf{B}$ is null, the partitioned matrix and its inverse become

$$\mathbf{P} = \begin{bmatrix} \mathbf{A} & \mathbf{0} \\ \mathbf{C} & \mathbf{D} \end{bmatrix} \qquad \mathbf{P}^{-1} = \begin{bmatrix} \mathbf{A}^{-1} & \mathbf{0} \\ -\mathbf{D}^{-1}\mathbf{CA}^{-1} & \mathbf{D}^{-1} \end{bmatrix} \tag{4-39}$$

as can be seen by substituting $\mathbf{B} = \mathbf{0}$ into Equations (4-37) or (4-38). The preceding formulas can be used for a lower triangular matrix, although of course they are not limited to such a matrix because $\mathbf{A}$ and $\mathbf{D}$ in Equation (4-39) do not have to be triangular. Similarly, if $\mathbf{C}$ is null, the expressions are

$$\mathbf{P} = \begin{bmatrix} \mathbf{A} & \mathbf{B} \\ \mathbf{0} & \mathbf{D} \end{bmatrix} \qquad \mathbf{P}^{-1} = \begin{bmatrix} \mathbf{A}^{-1} & -\mathbf{A}^{-1}\mathbf{BD}^{-1} \\ \mathbf{0} & \mathbf{D}^{-1} \end{bmatrix} \tag{4-40}$$

and these equations can be applied, for example, to an upper triangular matrix.

An even simpler situation arises when both off-diagonal submatrices are zero:

$$\mathbf{P} = \begin{bmatrix} \mathbf{A} & \mathbf{0} \\ \mathbf{0} & \mathbf{D} \end{bmatrix} \qquad \mathbf{P}^{-1} = \begin{bmatrix} \mathbf{A}^{-1} & \mathbf{0} \\ \mathbf{0} & \mathbf{D}^{-1} \end{bmatrix} \tag{4–41}$$

Inspection of the three preceding inversion formulas shows that they can be used only if both **A** and **D** have inverses. If either **A** or **D** is singular when one of the off-diagonal submatrices is null, the matrix **P** will also be singular (see Equations 3–13 and 3–14).

Example. Invert the following fourth-order matrix by partitioning:

$$\mathbf{P} = \left[\begin{array}{cc|cc} 2.847 & 1.721 & 1.638 & 0.843 \\ 2.012 & 0.945 & 4.237 & 1.658 \\ \hline 1.013 & 1.426 & 2.166 & 0.881 \\ 0.980 & 0.879 & 2.257 & 1.743 \end{array}\right]$$

Inasmuch as 2×2 matrices are easy to invert, the given matrix will be partitioned symmetrically, as shown above, so that the submatrices in Equation (4–35) are

$$\mathbf{A} = \begin{bmatrix} 2.847 & 1.721 \\ 2.012 & 0.945 \end{bmatrix} \qquad \mathbf{B} = \begin{bmatrix} 1.638 & 0.843 \\ 4.237 & 1.658 \end{bmatrix}$$

$$\mathbf{C} = \begin{bmatrix} 1.013 & 1.426 \\ 0.980 & 0.879 \end{bmatrix} \qquad \mathbf{D} = \begin{bmatrix} 2.166 & 0.881 \\ 2.257 & 1.743 \end{bmatrix}$$

Next, the submatrices **Q**, **R**, **S**, and **T** are obtained by direct substitution into Equations (4–37). The calculations proceed as follows, using the rule expressed by Equation (4–12) for finding the inverse of a 2×2 matrix:

$$\mathbf{D}^{-1} = \frac{1}{1.787}\begin{bmatrix} 1.743 & -0.881 \\ -2.257 & 2.166 \end{bmatrix} = \begin{bmatrix} 0.975 & -0.493 \\ -1.263 & 1.212 \end{bmatrix}$$

$$\mathbf{Q} = (\mathbf{A} - \mathbf{B}\mathbf{D}^{-1}\mathbf{C})^{-1} = \begin{bmatrix} 2.097 & 0.773 \\ 0.025 & -1.892 \end{bmatrix}^{-1} = \begin{bmatrix} 0.475 & 0.194 \\ 0.006 & -0.526 \end{bmatrix}$$

$$\mathbf{R} = -\mathbf{Q}\mathbf{B}\mathbf{D}^{-1} = \begin{bmatrix} -0.648 & -0.086 \\ 1.069 & -0.043 \end{bmatrix} \qquad \mathbf{S} = -\mathbf{D}^{-1}\mathbf{C}\mathbf{Q} = \begin{bmatrix} -0.246 & 0.406 \\ 0.048 & -0.369 \end{bmatrix}$$

$$\mathbf{T} = \mathbf{D}^{-1} - \mathbf{D}^{-1}\mathbf{C}\mathbf{R} = \begin{bmatrix} 0.279 & -0.408 \\ -0.536 & 1.173 \end{bmatrix}$$

In this calculation process, a total of two 2×2 matrices were inverted and six 2 X 2 multiplications were performed. The inverse matrix (see Equation 4-36) is

$$\mathbf{P}^{-1} = \begin{bmatrix} 0.475 & 0.194 & -0.648 & -0.086 \\ 0.006 & -0.526 & 1.069 & -0.043 \\ -0.246 & 0.406 & 0.279 & -0.408 \\ 0.048 & -0.369 & -0.536 & 1.173 \end{bmatrix}$$

which agrees with the result obtained in Example 5 of Section 4.7.

There are many other numerical schemes for matrix inversion, such as Gaussian elimination and factorization. Information on these methods can be found in the references listed at the end of the book. For small matrices that are to be inverted by hand, the methods described in Sections 4.7 and 4.8 are quite suitable. However, the inversion of large matrices requires the use of a digital computer in order to avoid tedious calculations and to guarantee accuracy in the results.

Problems

The problems for Section 4.4 are to be solved by using the adjoint matrix (see Equation 4-9). Verify the inverses by multiplying with the original matrices.

4.4-1. Find the inverse of each of the following matrices:

$$\mathbf{A}_1 = \begin{bmatrix} 4 & -2 \\ 1 & 2 \end{bmatrix} \qquad \mathbf{A}_2 = \begin{bmatrix} 3 & 1 \\ -5 & -2 \end{bmatrix} \qquad \mathbf{A}_3 = \begin{bmatrix} 7 & 12 \\ 3 & 2 \end{bmatrix}$$

4.4-2. Invert the matrix **A** shown below:

$$\mathbf{A} = \begin{bmatrix} 2 & 1 & 1 \\ 7 & 1 & 2 \\ 21 & 0 & 4 \end{bmatrix}$$

4.4-3. Invert the following matrix **B**:

$$\mathbf{B} = \begin{bmatrix} -4 & -19 & 11 \\ 1 & 5 & -2 \\ 2 & 7 & -12 \end{bmatrix}$$

4.4-4. Determine the inverse of the matrix shown:

$$\mathbf{C} = \begin{bmatrix} 2 & -1 & 3 \\ 0 & -2 & 4 \\ 5 & 1 & 6 \end{bmatrix}$$

4.4–5. Invert the matrix **D** shown below:

$$\mathbf{D} = \begin{bmatrix} 5 & 0 & 10 \\ -10 & 15 & 0 \\ -5 & 10 & 20 \end{bmatrix}$$

4.4–6. Find the inverse of the following symmetric matrix:

$$\mathbf{E} = \begin{bmatrix} 4 & 2 & -3 \\ 2 & 5 & 1 \\ -3 & 1 & 6 \end{bmatrix}$$

4.4–7. Find the inverse of the following matrix:

$$\mathbf{F} = \begin{bmatrix} 10 & 6 & 8 \\ 5 & -9 & -12 \\ 5 & 7 & -4 \end{bmatrix}$$

4.4–8. Find the inverse of the following lower triangular matrix:

$$\mathbf{G} = \begin{bmatrix} 1 & 0 & 0 \\ \lambda_1 & 1 & 0 \\ \lambda_2 & \lambda_3 & 1 \end{bmatrix}$$

4.4–9. Obtain the inverse of the fourth-order lower triangular matrix shown:

$$\mathbf{H} = \begin{bmatrix} 2 & 0 & 0 & 0 \\ 1 & -1 & 0 & 0 \\ 3 & -2 & 2 & 0 \\ 4 & 0 & 2 & 3 \end{bmatrix}$$

4.4–10. Determine the inverse of the following symmetric band matrix:

$$\mathbf{J} = \begin{bmatrix} 2 & 1 & 0 & 0 & 0 \\ 1 & 2 & 1 & 0 & 0 \\ 0 & 1 & 2 & 1 & 0 \\ 0 & 0 & 1 & 2 & 1 \\ 0 & 0 & 0 & 1 & 2 \end{bmatrix}$$

4.4–11. Determine the inverse of an nth order, lower triangular matrix having all nonzero elements equal to unity:

$$\mathbf{K} = \begin{bmatrix} 1 & 0 & 0 & \cdots & 0 & 0 \\ 1 & 1 & 0 & \cdots & 0 & 0 \\ 1 & 1 & 1 & \cdots & 0 & 0 \\ \cdots & \cdots & \cdots & \cdots & \cdots & \cdots \\ 1 & 1 & 1 & \cdots & 1 & 1 \end{bmatrix}$$

4.4–12. Obtain the inverse of the following matrix:

$$\mathbf{L} = \begin{bmatrix} S^2 & -SC & C^2 \\ -2SC & C^2-S^2 & 2SC \\ C^2 & SC & S^2 \end{bmatrix}$$

in which $S = \sin\alpha$ and $C = \cos\alpha$. (Note: The result may surprise you, because this matrix is neither symmetric nor orthogonal.)

4.5–1. Determine the inverses of the following matrices.

$$\mathbf{A} = \begin{bmatrix} 3 & 0 & 0 & 0 \\ 0 & 2 & 0 & 0 \\ 0 & 0 & -3 & 0 \\ 0 & 0 & 0 & 1 \end{bmatrix} \qquad \mathbf{B} = \begin{bmatrix} 0 & 0 & 0 & 2 \\ 0 & 0 & 4 & 0 \\ 0 & 3 & 0 & 0 \\ 2 & 0 & 0 & 0 \end{bmatrix}$$

4.5–2. Show that the inverse of a square matrix with nonzero terms only on the secondary diagonal is obtained by replacing each nonzero term with its reciprocal and then transposing the matrix.

4.5–3. Find the inverse of the skew-symmetric matrix **C** shown below:

$$\mathbf{C} = \begin{bmatrix} 0 & 2 & 0 & 1 \\ -2 & 0 & 1 & 0 \\ 0 & -1 & 0 & 4 \\ -1 & 0 & -4 & 0 \end{bmatrix}$$

4.5–4. Using the matrices $\mathbf{A}_1$, $\mathbf{A}_2$, and $\mathbf{A}_3$ from Problem 4.4–1, demonstrate that the inverse of a product is the product of the inverses in reverse order (see Equation 4–18).

4.5–5. Using the matrix **A** from Problem 4.4–2, demonstrate that the inverse of the transpose of a matrix is equal to the transpose of the inverse (see Equation 4–19).

4.5–6. Using the matrix **C** from Problem 4.4–4, demonstrate that the product of the determinants of a matrix and its inverse is unity (see Equation 4–21).

4.6–1. Using only the four numbers ±0.6 and ±0.8, construct eight different orthogonal matrices of second order, and find the determinant of each.

4.6–2. If **A** represents a 2×2 orthogonal matrix having elements A_{11}, A_{12}, A_{21}, and A_{22}, show that each of the following is also an orthogonal matrix:

$$\mathbf{A}_1 = \begin{bmatrix} A_{11} & A_{12} & 0 \\ A_{21} & A_{22} & 0 \\ 0 & 0 & 1 \end{bmatrix} \qquad \mathbf{A}_2 = \begin{bmatrix} \mathbf{A} & \mathbf{0} \\ \mathbf{0} & \mathbf{A} \end{bmatrix}$$

4.6–3. Show that the following matrix is orthogonal:

$$\mathbf{A}_3 = \begin{bmatrix} \cos\theta & \sin\theta & 0 \\ -\sin\theta & \cos\theta & 0 \\ 0 & 0 & 1 \end{bmatrix}$$

4.6–4. Determine the values of the elements A_{12} and A_{21} so that the following matrix is orthogonal:

$$\mathbf{A}_4 = \begin{bmatrix} 0.5 & A_{12} \\ A_{21} & 0.5 \end{bmatrix}$$

4.6–5. Is the following matrix orthogonal?

$$\mathbf{A}_5 = \begin{bmatrix} 0.482 & 0.609 & 0.630 \\ 0.609 & 0.284 & -0.741 \\ 0.630 & -0.741 & 0.234 \end{bmatrix}$$

4.6–6. Is the following matrix orthogonal?

$$\mathbf{A}_6 = \begin{bmatrix} 0.482 & 0.609 & 0.630 \\ 0.609 & -0.750 & 0.259 \\ -0.630 & -0.259 & 0.732 \end{bmatrix}$$

4.6–7. Find the values of x_1 and x_2 so that the following matrix is orthogonal:

$$\mathbf{A}_7 = \begin{bmatrix} 0.552 & 0.629 & x_2 \\ 0.629 & 0.117 & 0.769 \\ 0.547 & x_1 & -0.331 \end{bmatrix}$$

4.6–8. Demonstrate that the following symmetric orthogonal matrix is its own inverse:

$$\mathbf{A}_8 = \begin{bmatrix} S^2 & -SC & -SC & C^2 \\ -SC & -S^2 & C^2 & SC \\ -SC & C^2 & -S^2 & SC \\ C^2 & SC & SC & S^2 \end{bmatrix}$$

in which $S = \sin\alpha$ and $C = \cos\alpha$.

The problems for Section 4.7 are to be solved by the method of successive transformations. Verify the inverses by multiplying with the original matrices.

4.7–1. Invert each of the following matrices by operations on the rows:

$$\mathbf{A}_1 = \begin{bmatrix} 5 & -8 \\ -4 & 7 \end{bmatrix} \qquad \mathbf{A}_2 = \begin{bmatrix} 4 & 5 \\ 12 & 18 \end{bmatrix}$$

4.7–2. Invert each of the following matrices by operations on the columns:

$$\mathbf{B}_1 = \begin{bmatrix} 6 & -9 \\ 2 & 5 \end{bmatrix} \qquad \mathbf{B}_2 = \begin{bmatrix} -4 & 5 \\ -10 & 15 \end{bmatrix}$$

4.7-3. Invert each of the following matrices by operations on the rows:

$$\mathbf{C}_1 = \begin{bmatrix} 1.42 & 1.38 \\ 2.11 & 1.25 \end{bmatrix} \qquad \mathbf{C}_2 = \begin{bmatrix} 67.4 & -32.1 \\ 48.8 & 75.5 \end{bmatrix}$$

4.7-4. Invert each of the following matrices by operations on the columns:

$$\mathbf{D}_1 = \begin{bmatrix} 0.728 & 0.209 \\ 0.209 & 0.641 \end{bmatrix} \qquad \mathbf{D}_2 = \begin{bmatrix} 729 & -216 \\ -366 & 431 \end{bmatrix}$$

4.7-5. Invert the matrix **E** shown below by using row operations, and then check the result by recalculating the inverse using column operations:

$$\mathbf{E} = \begin{bmatrix} 2 & 1 & -4 \\ 3 & -5 & 2 \\ 0 & -1 & 1 \end{bmatrix}$$

4.7-6. Solve the preceding problem for the matrix **F** given below:

$$\mathbf{F} = \begin{bmatrix} 4 & -2 & -3 \\ 1 & 2 & -5 \\ -1 & 0 & 3 \end{bmatrix}$$

4.7-7. Invert the matrix **G** shown below using row operations:

$$\mathbf{G} = \begin{bmatrix} 2.91 & 1.76 & 0.83 \\ 1.53 & 2.62 & 1.60 \\ 0.66 & 1.25 & 2.39 \end{bmatrix}$$

4.7-8. Find the inverse of the following matrix using column operations:

$$\mathbf{H} = \begin{bmatrix} 18.2 & 11.1 & 7.2 \\ 9.5 & 24.6 & 8.7 \\ 8.7 & 6.3 & 19.8 \end{bmatrix}$$

4.7-9. Find the inverse of the matrix **J** using row operations:

$$\mathbf{J} = \begin{bmatrix} 182.3 & -97.2 & 43.9 \\ 112.4 & 207.0 & 60.2 \\ -63.7 & 55.4 & 89.8 \end{bmatrix}$$

4.7-10. Find the inverse of the matrix **J** in the preceding problem using column operations.

4.7-11. Obtain the inverse of the following matrix:

$$\mathbf{K} = \begin{bmatrix} 0.713 & 0.692 & -0.205 \\ 0.498 & -1.027 & 0.317 \\ 0.392 & 0.661 & -0.826 \end{bmatrix}$$

4.7-12. Determine the inverse of the following matrix:

$$\mathbf{L} = \begin{bmatrix} -0.635 & 0.407 & 0.657 \\ 0.407 & 0.899 & -0.163 \\ 0.657 & -0.163 & 0.736 \end{bmatrix}$$

4.7-13. Invert the fourth-order upper triangular matrix shown:

$$\mathbf{M} = \begin{bmatrix} 16.2 & -8.7 & 14.1 & 3.9 \\ 0 & 11.6 & -10.8 & 7.2 \\ 0 & 0 & 29.1 & 8.3 \\ 0 & 0 & 0 & 22.5 \end{bmatrix}$$

4.7-14. Find the inverse of the following lower triangular matrix:

$$\mathbf{N} = \begin{bmatrix} 0.218 & 0 & 0 & 0 \\ 0.472 & -0.361 & 0 & 0 \\ 0.225 & 0.708 & 1.163 & 0 \\ 0.107 & -0.285 & 0.699 & 0.804 \end{bmatrix}$$

4.7-15. Find the inverse of the following band-symmetric matrix:

$$\mathbf{P} = \begin{bmatrix} 4 & 1 & 0 & 0 & 0 \\ 1 & 2 & 1 & 0 & 0 \\ 0 & 1 & 2 & 1 & 0 \\ 0 & 0 & 1 & 2 & 1 \\ 0 & 0 & 0 & 1 & 4 \end{bmatrix}$$

4.7-16. Determine the inverse of the fourth-order matrix:

$$\mathbf{Q} = \begin{bmatrix} 2.746 & 1.903 & 0.784 & -0.617 \\ 2.078 & -3.910 & 1.821 & 0.907 \\ -0.446 & 0.682 & 1.364 & -0.883 \\ 0.512 & 0.341 & -0.631 & 1.092 \end{bmatrix}$$

4.7-17. Find the inverse of the following band-symmetric matrix:

$$\mathbf{R} = \begin{bmatrix} 1 & -1 & 0 & 0 & 0 & 0 \\ -1 & 1 & -1 & 0 & 0 & 0 \\ 0 & -1 & 1 & -1 & 0 & 0 \\ 0 & 0 & -1 & 1 & -1 & 0 \\ 0 & 0 & 0 & -1 & 1 & -1 \\ 0 & 0 & 0 & 0 & -1 & 1 \end{bmatrix}$$

4.8-1. Verify the expressions for **Q**, **R**, **S**, and **T** given in Equations (4-37) by forming the products $\mathbf{PP}^{-1}$ and $\mathbf{P}^{-1}\mathbf{P}$ (see Equations 4-35 and 4-36).

4.8-2. Repeat the preceding problem using the expressions for **Q**, **R**, **S**, and **T** given in Equations (4-38).

4.8-3. Determine the inverse of the following 3 X 3 matrix by using the method of partitioning:

$$\mathbf{A} = \begin{bmatrix} 20 & -14 & 3 \\ -2 & 1 & 0 \\ 5 & -4 & 1 \end{bmatrix}$$

(Note that this matrix was inverted previously in Example 2 of Section 4.7.)

4.8-4. Invert the 3 X 3 matrix **B** given in Problem 4.4-3 by partitioning.

4.8-5. Find the inverse of the following matrix by partitioning:

$$\mathbf{C} = \begin{bmatrix} -31.968 & 44.961 & -20.157 \\ 19.606 & -27.008 & 11.968 \\ -1.890 & 2.362 & -0.551 \end{bmatrix}$$

4.8-6. Obtain the inverse of the fourth-order lower triangular matrix **H** given in Problem 4.4-9 by partitioning.

4.8-7. Invert the following band-symmetric matrix by partitioning:

$$\mathbf{D} = \begin{bmatrix} 2 & 1 & 0 & 0 \\ 1 & 2 & 1 & 0 \\ 0 & 1 & 2 & 1 \\ 0 & 0 & 1 & 2 \end{bmatrix}$$

4.8-8. Determine the inverse of the 4 X 4 upper triangular matrix **M** given in Problem 4.7-13 by partitioning.

4.8-9. Calculate the inverse of the following matrix by partitioning:

$$\mathbf{E} = \begin{bmatrix} 1.782 & 0 & -0.071 & 0 \\ -0.630 & 0.803 & 1.109 & -0.803 \\ 0 & 1.243 & 1.917 & 0.375 \\ 0.712 & -0.790 & 0 & 0.906 \end{bmatrix}$$

4.8-10. Determine the inverse of the 4 X 4 matrix **Q** given in Problem 4.7-16 by partitioning.

5

...Simultaneous Equations...

5.1. Introduction. Simultaneous algebraic equations are encountered frequently in engineering problems. When the analysis of a physical system is based on a linear model, the simultaneous equations usually will be linear; that is, the unknown quantities appear only to the first power. Furthermore, the coefficients in the equations usually will be constants. Such a set of n equations relating n unknown quantities can be expressed in the general form

$$\begin{aligned} A_{11}X_1 + A_{12}X_2 + \ldots + A_{1n}X_n &= B_1 \\ A_{21}X_1 + A_{22}X_2 + \ldots + A_{2n}X_n &= B_2 \\ \ldots \quad \ldots \quad \ldots \quad \ldots &\quad \ldots \\ A_{n1}X_1 + A_{n2}X_2 + \ldots + A_{nn}X_n &= B_n \end{aligned} \tag{5-1}$$

in which each X is an unknown quantity, the A terms are the coefficients of the unknowns, and the B terms are the right-hand sides of the equations. The matrix representation of the above set of equations is

$$\mathbf{AX} = \mathbf{B} \tag{5-2}$$

in which **A** is the matrix of coefficients, **X** is the vector of unknowns, and **B** is the vector of right-hand sides, as follows:

$$\mathbf{A} = \begin{bmatrix} A_{11} & A_{12} & \ldots & A_{1n} \\ A_{21} & A_{22} & \ldots & A_{2n} \\ \ldots & \ldots & \ldots & \ldots \\ A_{n1} & A_{n2} & \ldots & A_{nn} \end{bmatrix} \qquad \mathbf{X} = \begin{bmatrix} X_1 \\ X_2 \\ \ldots \\ X_n \end{bmatrix} \qquad \mathbf{B} = \begin{bmatrix} B_1 \\ B_2 \\ \ldots \\ B_n \end{bmatrix}$$

In this illustration the matrix **A** is square and of order n, and the vectors **X** and **B** are matrices of order $n \times 1$.

It is assumed for the present that Equations (5-1) are solvable for the n unknown quantities, as is expected in most physical problems. In a later section (Section 5.8) the criteria for the existence of a solution will be discussed. Also, it is assumed in the next three sections (which discuss methods of solving simultaneous equations) that the vector **B** is not a null vector. Equations of this form are called *nonhomogeneous equations*. If **B** is null, the equations are *homogeneous equations*, and their solution is described in Sections 5.5 and 5.8.

5.2. Solution by Inversion. A direct means of solving simultaneous equations involves the use of the inverse of the matrix of coefficients. If both sides of Equation (5-2) are premultiplied by the inverse of **A**, the result is

$$\mathbf{A}^{-1}\mathbf{A}\mathbf{X} = \mathbf{A}^{-1}\mathbf{B}$$

Therefore, the vector **X** of unknowns is

$$\mathbf{X} = \mathbf{A}^{-1}\mathbf{B} \tag{5-3}$$

This equation shows that **X** can be obtained by premultiplying the vector **B** by the inverse of **A**. The equation also shows the close relationship between matrix inversion and the solution of simultaneous equations. A unique solution of the equations will exist only if the inverse of the matrix of coefficients exists, which means that **A** must be nonsingular. If **A** is singular, there will be either no solution or an infinite number of solutions, as discussed in Section 5.8.

To illustrate the solution of simultaneous equations by inversion, consider the following three equations:

$$\begin{aligned} 2X_1 + 4X_2 + X_3 &= -11 \\ -X_1 + 3X_2 - 2X_3 &= -16 \\ 2X_1 - 3X_2 + 5X_3 &= 21 \end{aligned} \tag{5-4}$$

For these equations the matrix **A** of the coefficients and the vector **B** of right-hand sides are, respectively,

$$\mathbf{A} = \begin{bmatrix} 2 & 4 & 1 \\ -1 & 3 & -2 \\ 2 & -3 & 5 \end{bmatrix} \qquad \mathbf{B} = \begin{bmatrix} -11 \\ -16 \\ 21 \end{bmatrix}$$

The inverse of **A**, which can be found by one of the methods described in the preceding chapter, is equal to

$$\mathbf{A}^{-1} = \frac{1}{19}\begin{bmatrix} 9 & -23 & -11 \\ 1 & 8 & 3 \\ -3 & 14 & 10 \end{bmatrix}$$

Taking the product $\mathbf{A}^{-1}\mathbf{B}$ gives, for the vector of unknowns,

$$\mathbf{X} = \begin{bmatrix} X_1 \\ X_2 \\ X_3 \end{bmatrix} = \frac{1}{19}\begin{bmatrix} 9 & -23 & -11 \\ 1 & 8 & 3 \\ -3 & 14 & 10 \end{bmatrix}\begin{bmatrix} -11 \\ -16 \\ 21 \end{bmatrix} = \begin{bmatrix} 2 \\ -4 \\ 1 \end{bmatrix}$$

from which $X_1 = 2$, $X_2 = -4$, and $X_3 = 1$. Thus, the solution of the simultaneous equations has been obtained by using the inverse of the matrix of coefficients. The final result may be checked by performing the multiplication **AX** and observing that this product is equal to the matrix **B**.

The preceding method of solution is particularly advantageous if a set of equations is to be solved for several different vectors of right-hand sides. Once the inverse of the matrix of coefficients is determined, it is a relatively simple matter to calculate several vectors of unknowns corresponding to several different vectors **B**. This procedure is more efficient than attempting to resolve the entire set of equations for each different vector of right-hand sides, as is necessary when solving the equations by other methods, such as the method of elimination (described later in Section 5.4).

There are some occasions, however, when it may be convenient to solve the simultaneous equations by a direct solution for the unknowns without explicitly obtaining the inverse of the coefficient matrix. Many techniques are available for this purpose, and two such methods will be described in the next sections. Additional methods can be found in the references.

5.3. Cramer's Rule. A method for solving simultaneous equations by the use of determinants, known as *Cramer's rule*, can be derived directly from Equation (5-3) and the definition of the matrix inverse in terms of its adjoint. Substitution into Equation (5-3) of the expression for the inverse of a matrix, as given by Equation (4-9), yields the following expression for the vector of unknowns:

$$\mathbf{X} = \frac{\mathbf{A}^a}{|\mathbf{A}|}\mathbf{B}$$

in which $\mathbf{A}^a$ is the adjoint of the coefficient matrix **A**. Again assuming that there are n simultaneous equations, and also writing the adjoint in terms of cofactors of the determinant $|\mathbf{A}|$ (see Equation 4-8), it can be seen that the preceding equation for **X** becomes

$$\mathbf{X} = \frac{1}{|\mathbf{A}|}\begin{bmatrix} A_{11}^c & A_{21}^c & \dots & A_{n1}^c \\ A_{12}^c & A_{22}^c & \dots & A_{n2}^c \\ \dots & \dots & \dots & \dots \\ A_{1n}^c & A_{2n}^c & \dots & A_{nn}^c \end{bmatrix}\begin{bmatrix} B_1 \\ B_2 \\ \dots \\ B_n \end{bmatrix} \qquad (5\text{-}5)$$

In the above expression the adjoint matrix is of order $n \times n$, the determinant $|\mathbf{A}|$ is of order $n \times n$, and $\mathbf{X}$ and $\mathbf{B}$ are of order $n \times 1$. When the product of the adjoint matrix and the $\mathbf{B}$ matrix is formed, the equation for the vector $\mathbf{X}$ becomes

$$\mathbf{X} = \begin{bmatrix} X_1 \\ X_2 \\ \ldots \\ X_n \end{bmatrix} = \frac{1}{|\mathbf{A}|} \begin{bmatrix} B_1A_{11}^c + B_2A_{21}^c + \ldots + B_nA_{n1}^c \\ B_1A_{12}^c + B_2A_{22}^c + \ldots + B_nA_{n2}^c \\ \ldots \quad \ldots \quad \ldots \quad \ldots \\ B_1A_{1n}^c + B_2A_{2n}^c + \ldots + B_nA_{nn}^c \end{bmatrix} \tag{5-6}$$

Equating individual elements from the left- and right-hand sides of Equation (5-6) gives the following set of relations for the n unknowns:

$$\begin{aligned} X_1 &= \frac{B_1A_{11}^c + B_2A_{21}^c + \ldots + B_nA_{n1}^c}{|\mathbf{A}|} \\ X_2 &= \frac{B_1A_{12}^c + B_2A_{22}^c + \ldots + B_nA_{n2}^c}{|\mathbf{A}|} \\ \ldots &\quad \ldots \\ X_n &= \frac{B_1A_{1n}^c + B_2A_{2n}^c + \ldots + B_nA_{nn}^c}{|\mathbf{A}|} \end{aligned} \tag{5-7}$$

Inspection of the numerator in each of these expressions shows that it represents the expansion of a determinant by cofactors. For example, the numerator in the expression for X_1 is equal to the expansion of the following determinant by cofactors of the first column:

$$\begin{vmatrix} B_1 & A_{12} & A_{13} & \ldots & A_{1n} \\ B_2 & A_{22} & A_{23} & \ldots & A_{2n} \\ \ldots & \ldots & \ldots & \ldots & \ldots \\ B_n & A_{n2} & A_{n3} & \ldots & A_{nn} \end{vmatrix}$$

In a similar way, each of the other numerators can be written as a determinant, the difference being that the elements of $\mathbf{B}$ appear each time in a different column of the determinant. Therefore, Equations (5-7) can be written in the form

$$X_1 = \frac{\begin{vmatrix} B_1 & A_{12} & A_{13} & \ldots & A_{1n} \\ B_2 & A_{22} & A_{23} & \ldots & A_{2n} \\ \ldots & \ldots & \ldots & \ldots & \ldots \\ B_n & A_{n2} & A_{n3} & \ldots & A_{nn} \end{vmatrix}}{|\mathbf{A}|} \tag{5-8a}$$

$$X_2 = \frac{\begin{vmatrix} A_{11} & B_1 & A_{13} & \ldots & A_{1n} \\ A_{21} & B_2 & A_{23} & \ldots & A_{2n} \\ \ldots & \ldots & \ldots & \ldots & \ldots \\ A_{n1} & B_n & A_{n3} & \ldots & A_{nn} \end{vmatrix}}{|\mathbf{A}|} \tag{5-8b}$$

and so on until the last expression is obtained:

$$X_n = \frac{\begin{vmatrix} A_{11} & A_{12} & A_{13} & \ldots & B_1 \\ A_{21} & A_{22} & A_{23} & \ldots & B_2 \\ \ldots & \ldots & \ldots & \ldots & \ldots \\ A_{n1} & A_{n2} & A_{n3} & \ldots & B_n \end{vmatrix}}{|\mathbf{A}|} \tag{5-8c}$$

In the denominator of each expression appears the determinant of **A**, and in the numerator appears that same determinant modified by the replacement of one column with the elements from the **B** matrix. The column that has been replaced corresponds to the number of the unknown itself; thus, the first column has been replaced in the expression for X_1; the second column has been replaced in the expression for X_2; and so on for all n unknowns.

The result derived above is known as Cramer's rule,* which can be stated in the following way. The solution for any one of the unknowns X_i in a set of simultaneous equations is equal to the ratio of two determinants; the determinant in the denominator is the determinant of the coefficients, while the determinant in the numerator is that same determinant with the ith column replaced by the elements from the right-hand sides of the equations.

If Cramer's rule is applied to the solution of the three equations given previously (see Equations 5-4), the expression for X_1 becomes

$$X_1 = \frac{\begin{vmatrix} -11 & 4 & 1 \\ -16 & 3 & -2 \\ 21 & -3 & 5 \end{vmatrix}}{\begin{vmatrix} 2 & 4 & 1 \\ -1 & 3 & -2 \\ 2 & -3 & 5 \end{vmatrix}} = \frac{38}{19} = 2$$

which agrees with the result obtained before. In a similar manner the remaining unknowns can be found:

*Gabriel Cramer (1704–1752) was a Swiss mathematician.

$$X_2 = \frac{\begin{vmatrix} 2 & -11 & 1 \\ -1 & -16 & -2 \\ 2 & 21 & 5 \end{vmatrix}}{\begin{vmatrix} 2 & 4 & 1 \\ -1 & 3 & -2 \\ 2 & -3 & 5 \end{vmatrix}} = -\frac{76}{19} = -4$$

$$X_3 = \frac{\begin{vmatrix} 2 & 4 & -11 \\ -1 & 3 & -16 \\ 2 & -3 & 21 \end{vmatrix}}{\begin{vmatrix} 2 & 4 & 1 \\ -1 & 3 & -2 \\ 2 & -3 & 5 \end{vmatrix}} = \frac{19}{19} = 1$$

Cramer's rule is useful in hand calculations only if the determinants can be evaluated easily, as in the above example. When the number of equations is large, other methods of solution are more desirable.

5.4. Method of Elimination. One of the most widely used methods for solving simultaneous equations is the *method of elimination* (also called *Gaussian elimination**). The use of the term elimination denotes the fact that the unknowns are eliminated successively from the equations. For example, if the original equations are in the form shown by Equations (5-1), it is possible to begin by solving the first equation for X_1 in terms of the other unknowns. Then this expression for X_1 is substituted into the remaining $n - 1$ equations, giving a set of equations with $n - 1$ unknown quantities ($X_2, X_3, \ldots, X_n$). The process may now be repeated by solving for X_2 in terms of $X_3, X_4, \ldots, X_n$ from the first of the new equations, and then substituting this relation into the remaining $n - 2$ equations. The result is a set of $n - 2$ equations containing $n - 2$ unknowns. This process may be continued until finally the last equation, containing only one unknown, is obtained. When the value of X_n is found from the last equation, it can be substituted into the preceding equation (which contains X_{n-1} and X_n as unknowns), and hence X_{n-1} can be found. Next, these two values are substituted back into the preceding equation, which then is solved for X_{n-2}. The process is continued until the first equation is reached, and the last unknown X_1 is calculated.

The method of elimination can be illustrated by again solving Equations (5-4), which are repeated as follows:

*Carl Friedrich Gauss (1777–1855) was a German mathematician and scientist.

$$\begin{aligned} 2X_1 + 4X_2 + X_3 &= -11 \\ -X_1 + 3X_2 - 2X_3 &= -16 \\ 2X_1 - 3X_2 + 5X_3 &= 21 \end{aligned} \qquad \text{(5-4) repeated}$$

Solving the first equation for X_1 in terms of X_2 and X_3 gives

$$X_1 = -2X_2 - 0.5X_3 - 5.5 \qquad \text{(a)}$$

When this expression is substituted into the last two of Equations (5-4), they become

$$\begin{aligned} 5X_2 - 1.5X_3 &= -21.5 \\ -7X_2 + 4X_3 &= 32 \end{aligned} \qquad \text{(b)}$$

Next, the first of Equations (b) is solved for X_2 in terms of X_3:

$$X_2 = 0.3X_3 - 4.3 \qquad \text{(c)}$$

and then this expression for X_2 is substituted into the second of Equations (b), which in turn can be solved for X_3, giving

$$X_3 = 1$$

Substitution of this value for X_3 into the equation for X_2 in terms of X_3 (Equation c) makes it possible to solve for X_2:

$$X_2 = -4$$

Lastly, the values for X_2 and X_3 are substituted into the equation for X_1 (Equation a) yielding the result

$$X_1 = 2$$

and, therefore, all the unknowns have been calculated. The same general scheme of solution can be used when there are any number of simultaneous equations to be solved.

Various modifications have been made to the method of elimination. One of these is the *Gauss-Jordan method*,* in which the expression for each unknown X_i is substituted not only into the equations that follow the one from which it was obtained but also into all of the earlier expressions. Then, when the last equation is solved for X_n, all of the other unknowns can be found immediately by direct substitution. For example, in the numerical problem of the preceding paragraph, after solving for X_2 and obtaining (see Equation c)

$$X_2 = 0.3X_3 - 4.3$$

*Camille Jordan (1838–1922) was a French mathematician.

one would next substitute back into the expression for X_1 (Equation a) to obtain

$$X_1 = -1.1X_3 + 3.1$$

This substitution means that both X_1 and X_2 have been expressed in terms of the single unknown X_3, after which the solution continues in the same manner as before. Upon solving for X_3, it becomes possible to find both X_1 and X_2 by a single substitution for each.

In practical work it is important to make the calculations as systematic as possible and to reduce errors caused by rounding off the numbers. The latter result can be accomplished by another version of the method of elimination. In this approach the equation having the largest coefficient of X_1 is placed first; that is, this equation is taken as the leading or *pivotal equation*. The first step in the solution is to divide the entire first equation by the coefficient of the first term, which is called the pivot element. Then the equation that remains is multiplied by the coefficient of X_1 from the second equation. This new equation is subtracted from the second equation in order to eliminate the first term from that equation. The same steps are followed for the remainder of the equations, giving a set of $n - 1$ equations with $X_2, X_3, \ldots, X_n$ as unknowns. Then the process is repeated for the new set of $n - 1$ equations, with the equation having the largest coefficient of X_2 taken as the leading equation. Eventually the last unknown X_n can be found, whereupon the other unknowns can be found through back substitution. This method is called Gaussian elimination using the largest pivots.

Again using Equations (5-4) for illustration, it can be noted that the equations are already in satisfactory order, because the first equation has a coefficent for X_1 which is as large as the coefficient in either of the remaining equations. First, the leading equation is divided by the coefficient from the pivotal term to obtain

$$X_1 + 2X_2 + 0.5X_3 = -5.5 \tag{d}$$

Equation (d) then is added to the second equation, giving the result

$$5X_2 - 1.5X_3 = -21.5 \tag{e}$$

Also, two times Equation (d) is subtracted from the third equation:

$$-7X_2 + 4X_3 = 32 \tag{f}$$

Next, Equations (e) and (f) are treated in a similar manner. The equations are interchanged so that Equation (f) becomes the leading equation:

$$\begin{aligned} -7X_2 + 4X_3 &= 32 \\ 5X_2 - 1.5X_3 &= -21.5 \end{aligned} \tag{g}$$

The first of these equations is divided by -7:

$$X_2 - \frac{4}{7}X_3 = -\frac{32}{7}$$

and then five times this equation is subtracted from the second of Equations (g):

$$\frac{19}{14}X_3 = \frac{19}{14}$$

from which $X_3 = 1$. By the process of back substitution, it is found that $X_2 = -4$ and $X_1 = 2$.

The reader probably has observed already the similarities between the method of elimination for solving simultaneous equations, the method of successive transformations for determining the inverse of a matrix, and the method of pivotal condensation for evaluating a determinant. All of these methods are related to one another and may be considered to be variations of the method of elimination.

5.5. Homogeneous Equations. A set of simultaneous equations is homogeneous if all of the constant terms on the right-hand sides of the equal signs are zero. Thus, the general form of n homogeneous equations involving n unknown quantities is

$$\begin{aligned} A_{11}X_1 + A_{12}X_2 + \ldots + A_{1n}X_n &= 0 \\ A_{21}X_1 + A_{22}X_2 + \ldots + A_{2n}X_n &= 0 \\ \ldots \quad \ldots \quad \ldots \quad \ldots & \\ A_{n1}X_1 + A_{n2}X_2 + \ldots + A_{nn}X_n &= 0 \end{aligned} \tag{5-9}$$

or, in matrix form,

$$\mathbf{AX} = \mathbf{0} \tag{5-10}$$

in which **A** is a square matrix of coefficients, **X** is the vector of unknowns, and **0** is a null vector.

It can be seen by inspection of Equations (5-9) that one solution of the equations is

$$X_1 = X_2 = \ldots = X_n = 0$$

or

$$\mathbf{X} = \mathbf{0} \tag{5-11}$$

Such a solution, in which all of the unknowns are zero, is called the *trivial* solution. Of course, it is usually nontrivial solutions that are of interest in physical problems.

A nontrivial solution to Equation (5-10) occurs only when the matrix **A** of coefficients is a singular matrix. This conclusion follows from Cramer's rule (see Equations 5-8), which expresses the solution for each of the unknown quantities as the ratio of two determinants. When the equations are homogeneous, the determinants appearing in the numerators all have the value zero because one column (the column containing $B_1, B_2, \ldots, B_n$) consists of zero elements. Thus, Equations (5-8) become

$$X_1 = \frac{0}{|\mathbf{A}|} \qquad X_2 = \frac{0}{|\mathbf{A}|} \qquad \cdots \qquad X_n = \frac{0}{|\mathbf{A}|}$$

and the solution vector **X** becomes

$$\mathbf{X} = \frac{\mathbf{0}}{|\mathbf{A}|} \tag{5-12}$$

in which **0** is a null vector of the same order as **X**. Now, if the determinant of **A** has any value other than zero, it follows that the vector **X** must be null (the trivial solution). The only possibility for a nontrivial solution is when the determinant of **A** is zero, which means that the matrix **A** is singular. Actually, it is possible to prove that when **A** is singular there always will be a nontrivial solution; that is, the fact that the determinant of **A** is equal to zero is both a necessary and sufficient condition for the existence of a nontrivial solution. Furthermore, if one nontrivial solution exists, then an infinite number of such solutions will exist. This conclusion can be seen by noting that if a nonzero vector **X** is a solution of Equation (5-10), then $\beta\mathbf{X}$ is also a solution, where β is any arbitrary constant.

Additional considerations relating to the solution of homogeneous equations require a study of linear dependence of vectors and the rank of a matrix; therefore, further discussion of such equations is postponed until Section 5.8. For present purposes, a few simple examples will serve to illustrate the solution of homogeneous equations.

Example 1. Obtain the solution of the following simultaneous equations:

$$2X_1 - 7X_2 = 0$$

$$4X_1 - 9X_2 = 0$$

Since the determinant of the coefficient matrix **A** has the value

$$|\mathbf{A}| = \begin{vmatrix} 2 & -7 \\ 4 & -9 \end{vmatrix} = 2(-9) - (-7)(4) = 10$$

it follows that the matrix **A** is not singular; hence, the only solution is $X_1 = X_2 = 0$, or $\mathbf{X} = \mathbf{0}$.

Example 2. Obtain the solution of the following simultaneous equations:

$$3X_1 + 5X_2 = 0$$

$$9X_1 + 15X_2 = 0$$

In this case, evaluation of the determinant of **A** shows that it is singular; therefore, nontrivial solutions exist. From either of the two equations the following relationship between X_1 and X_2 is found:

$$X_1 = -\frac{5}{3}X_2$$

Thus, for any assumed value of one of the unknowns, the other unknown can be obtained. This result can be expressed in another form by stating the vector **X** itself; for instance:

$$\mathbf{X} = \beta \begin{bmatrix} -5 \\ 3 \end{bmatrix}$$

in which β is a constant that can have any value (including zero).

Example 3. Determine the solution of the following simultaneous equations:

$$X_1 + X_2 - X_3 = 0$$

$$-4X_1 + 5X_2 + 7X_3 = 0$$

$$2X_1 - X_2 - 3X_3 = 0$$

The coefficient matrix **A** is singular, and hence nontrivial solutions exist. Solving the equations by the method of elimination gives

$$X_1 = \frac{4}{3}X_3 \qquad X_2 = -\frac{1}{3}X_3$$

Thus, for any assumed value of X_3, the other unknowns (X_1 and X_2) are determined and the vector **X** may be expressed as

$$\mathbf{X} = \beta \begin{bmatrix} 4 \\ -1 \\ 3 \end{bmatrix}$$

in which β can have any value (including zero).

In the preceding two examples the solutions contain only the ratios of the unknown quantities. Hence, the magnitudes of the unknowns are not determined; instead, they can be adjusted arbitrarily by assigning various values to the constant β. It will be shown later in Section 5.8 that other sets of homogeneous equations have more complicated solutions involving more than one arbitrary constant.

5.6. Linear Dependence of Vectors. In order to understand more fully the nature of the solutions of simultaneous equations, some additional concepts are needed. For this purpose, linear dependence of vectors is discussed in this section, and the rank of a matrix is discussed in Section 5.7.

Consider a set of either row or column vectors $(\mathbf{A}_1, \mathbf{A}_2, \ldots, \mathbf{A}_n)$ such that each vector in the set has the same dimension. The set of vectors is said to be a *linearly dependent* set, and the vectors in the set are said to be *linearly dependent*, if there exists a relationship among the vectors of the following form:

$$\alpha_1 \mathbf{A}_1 + \alpha_2 \mathbf{A}_2 + \ldots + \alpha_n \mathbf{A}_n = 0 \tag{5-13}$$

in which $\alpha_1, \alpha_2, \ldots, \alpha_n$ are scalar numbers such that at least one of them is not zero. If no such relationship exists, the vectors are *linearly independent*.

As an example of linear dependence or independence, consider the following set of four-dimensional column vectors:

$$\begin{aligned} \mathbf{A}_1 &= \{2 \quad -6 \quad 8 \quad -2\} \\ \mathbf{A}_2 &= \{3 \quad -9 \quad 12 \quad -3\} \\ \mathbf{A}_3 &= \{5 \quad 3 \quad -2 \quad 1\} \\ \mathbf{A}_4 &= \{2 \quad 5 \quad -1 \quad 4\} \end{aligned} \tag{a}$$

Equation (5-13) will be used to establish the linear dependence or independence of these vectors. Substitution of the four vectors into Equation (5-13) produces the following expression:

$$\alpha_1 \begin{bmatrix} 2 \\ -6 \\ 8 \\ -2 \end{bmatrix} + \alpha_2 \begin{bmatrix} 3 \\ -9 \\ 12 \\ -3 \end{bmatrix} + \alpha_3 \begin{bmatrix} 5 \\ 3 \\ -2 \\ 1 \end{bmatrix} + \alpha_4 \begin{bmatrix} 2 \\ 5 \\ -1 \\ 4 \end{bmatrix} = \begin{bmatrix} 0 \\ 0 \\ 0 \\ 0 \end{bmatrix} \tag{b}$$

Equation (b) is equivalent to four homogeneous equations having α_1, α_2, α_3, and α_4 as unknowns:

$$\begin{aligned} 2\alpha_1 + 3\alpha_2 + 5\alpha_3 + 2\alpha_4 &= 0 \\ -6\alpha_1 - 9\alpha_2 + 3\alpha_3 + 5\alpha_4 &= 0 \\ 8\alpha_1 + 12\alpha_2 - 2\alpha_3 - \alpha_4 &= 0 \\ -2\alpha_1 - 3\alpha_2 + \alpha_3 + 4\alpha_4 &= 0 \end{aligned} \tag{c}$$

When these equations are solved by the method of elimination, as demonstrated in Example 3 of the preceding section, the following results are obtained:

$$\begin{bmatrix} \alpha_1 \\ \alpha_2 \\ \alpha_3 \\ \alpha_4 \end{bmatrix} = \beta \begin{bmatrix} 3 \\ -2 \\ 0 \\ 0 \end{bmatrix}$$

Therefore, it may be concluded that the vectors $\mathbf{A}_1$, $\mathbf{A}_2$, $\mathbf{A}_3$, and $\mathbf{A}_4$ (see Equations a) satisfy the following equation:

$$3\mathbf{A}_1 - 2\mathbf{A}_2 + (0)\mathbf{A}_3 + (0)\mathbf{A}_4 = \mathbf{0} \tag{d}$$

or any scalar multiple thereof. Consequently, the four vectors constitute a linearly dependent set. (If the only solution to Equations (c) had been the trivial solution, then the vectors would have been linearly independent.)

The two vectors $\mathbf{A}_1$ and $\mathbf{A}_2$ from Equations (a) also are a linearly dependent set, since

$$3\mathbf{A}_1 - 2\mathbf{A}_2 = \mathbf{0} \tag{e}$$

as can be seen by inspection of Equation (d). However, the vectors $\mathbf{A}_2$, $\mathbf{A}_3$, and $\mathbf{A}_4$ are a linearly independent set because the equation

$$\alpha_2 \mathbf{A}_2 + \alpha_3 \mathbf{A}_3 + \alpha_4 \mathbf{A}_4 = \mathbf{0} \tag{f}$$

is true only if α_2, α_3, and α_4 are all zero. This conclusion can be confirmed by substituting vectors $\mathbf{A}_2$, $\mathbf{A}_3$, and $\mathbf{A}_4$ into Equation (5–13), which leads to a set of four homogeneous equations similar to Equations (c) but having only α_2, α_3, and α_4 as unknowns:

$$\begin{aligned} 3\alpha_2 + 5\alpha_3 + 2\alpha_4 &= 0 \\ -9\alpha_2 + 3\alpha_3 + 5\alpha_4 &= 0 \\ 12\alpha_2 - 2\alpha_3 - \alpha_4 &= 0 \\ -3\alpha_2 + \alpha_3 + 4\alpha_4 &= 0 \end{aligned} \tag{g}$$

When Equations (g) are solved by the method of elimination, it is found that their only solution is

$$\alpha_2 = \alpha_3 = \alpha_4 = 0$$

as stated above. Therefore, the set of vectors $\mathbf{A}_2$, $\mathbf{A}_3$, and $\mathbf{A}_4$ is a linearly independent set.

If a set of vectors is linearly independent, then any subset of those vectors is also linearly independent. This idea can be seen from the preceding example involving the set of vectors $\mathbf{A}_2$, $\mathbf{A}_3$, and $\mathbf{A}_4$. Since these vectors satisfy Equation (f) only when α_2, α_3, and α_4 are all zero, it can be seen immediately that there is no possibility for any two of them to satisfy Equation (5–13) except when the α's are zero. For example, the equation

$$\alpha_2 \mathbf{A}_2 + \alpha_3 \mathbf{A}_3 = \mathbf{0}$$

will be satisfied only if α_2 and α_3 are zero. Hence, the six sets of vectors $\mathbf{A}_2$ and $\mathbf{A}_3$; $\mathbf{A}_2$ and $\mathbf{A}_4$; $\mathbf{A}_3$ and $\mathbf{A}_4$; $\mathbf{A}_2$; $\mathbf{A}_3$; and $\mathbf{A}_4$ are each linearly independent because they are subsets of a set of vectors that is known to be linearly independent.

On the other hand, if a set of vectors (such as $\mathbf{A}_1$ and $\mathbf{A}_2$; see Equation e) is linearly dependent, any larger set consisting of vectors having the same dimension and containing the original set of vectors is also linearly dependent. To demonstrate this conclusion, it is only necessary to observe that the values of α appearing in Equation (5-13) may be taken as zero for the additional vectors. Thus, inasmuch as $\mathbf{A}_1$ and $\mathbf{A}_2$ in the above example are linearly dependent, it is also true that the following three sets of vectors are linearly dependent: $\mathbf{A}_1$, $\mathbf{A}_2$, and $\mathbf{A}_3$; $\mathbf{A}_1$, $\mathbf{A}_2$, and $\mathbf{A}_4$; and $\mathbf{A}_1$, $\mathbf{A}_2$, $\mathbf{A}_3$, and $\mathbf{A}_4$. Furthermore, any of these sets of vectors can be augmented by as many more four-dimensional vectors as desired, and the resulting set of vectors will be linearly dependent also.

If the set of vectors contains only one vector, then for linear dependence it is necessary that

$$\alpha_1 \mathbf{A}_1 = \mathbf{0}$$

in which α_1 is not zero. Under these conditions the equation will be satisfied only if $\mathbf{A}_1$ is null. It can be concluded, therefore, that a single vector is linearly dependent only if it is a null vector; otherwise it is linearly independent. If a set of vectors contains a null vector, the vectors are linearly dependent, inasmuch as the null vector itself is linearly dependent.

If a vector $\mathbf{A}$ can be expressed in terms of other vectors by an equation of the form

$$\mathbf{A} = \gamma_1 \mathbf{A}_1 + \gamma_2 \mathbf{A}_2 + \cdots + \gamma_n \mathbf{A}_n \tag{5-14}$$

in which $\gamma_1, \gamma_2, \ldots, \gamma_n$ are any scalar numbers (including zero), then the vector $\mathbf{A}$ is said to be a *linear combination* of the other vectors. For instance, the vector $\mathbf{A}$, defined as

$$\mathbf{A} = \{-9 \quad -16 \quad 21 \quad -3\}$$

is a linear combination of the vectors $\mathbf{A}_1$, $\mathbf{A}_2$, $\mathbf{A}_3$, and $\mathbf{A}_4$ given in the previous example, inasmuch as

$$\mathbf{A} = 2\mathbf{A}_1 + (0)\mathbf{A}_2 - 3\mathbf{A}_3 + \mathbf{A}_4$$

Of course, the vector $\mathbf{A}$ is also a linear combination of the vectors $\mathbf{A}_1$, $\mathbf{A}_3$, and $\mathbf{A}_4$, but it is not expressible, for example, as a linear combination of the vectors $\mathbf{A}_1$ and $\mathbf{A}_2$ by themselves. When a vector is expressible as a linear combination of other vectors, it is said to be *linearly dependent* upon those vectors. Thus, the vector $\mathbf{A}$ in the above example is linearly dependent upon $\mathbf{A}_1$, $\mathbf{A}_3$, and $\mathbf{A}_4$, but is linearly independent of $\mathbf{A}_1$ and $\mathbf{A}_2$.

Consider now a set of n linearly dependent vectors as defined by Equation

(5–13). Since at least one of the values of α, say α_i, is not zero, each term in the equation can be divided by that value; upon rearranging, the result is

$$\mathbf{A}_i = -\frac{\alpha_1}{\alpha_i}\mathbf{A}_1 - \frac{\alpha_2}{\alpha_i}\mathbf{A}_2 - \cdots - \frac{\alpha_n}{\alpha_i}\mathbf{A}_n \tag{5-15}$$

This equation shows that each vector $\mathbf{A}_i$ (having a nonzero scalar multiplier α_i) in a set of linearly dependent vectors can be expressed as a linear combination of the other vectors (compare Equations 5-15 and 5-14). It follows also that if one vector in a set of vectors can be expressed as a linear combination of the others, the vectors are linearly dependent; if no vector can be expressed as a linear combination, the vectors are linearly independent.

As mentioned earlier, the concept of linear dependence discussed in this section applies to sets of either row or column vectors (both cases are illustrated in the examples that follow). Furthermore, the ideas can be extended by analogy to other mathematical entities, such as polynomials and matrices; however, only the notion of linear dependence of vectors will be required in later discussions.

Example 1. Are the column vectors $\mathbf{A}_1$, $\mathbf{A}_2$, and $\mathbf{A}_3$ (see below) linearly dependent or independent?

$$\mathbf{A}_1 = \{2 \quad -1 \quad 3\} \qquad \mathbf{A}_2 = \{-3 \quad 1 \quad 2\} \qquad \mathbf{A}_3 = \{5 \quad 0 \quad 2\}$$

If the vectors are linearly dependent, they will satisfy Equation (5–13), which leads to the following simultaneous equations:

$$\begin{aligned} 2\alpha_1 - 3\alpha_2 + 5\alpha_3 &= 0 \\ -\alpha_1 + \alpha_2 \qquad &= 0 \\ 3\alpha_1 + 2\alpha_2 + 2\alpha_3 &= 0 \end{aligned}$$

These equations are homogeneous and will have nontrivial solutions only if the coefficient matrix is singular. The determinant of the coefficients is

$$\begin{vmatrix} 2 & -3 & 5 \\ -1 & 1 & 0 \\ 3 & 2 & 2 \end{vmatrix} = -27$$

Therefore, the matrix is not singular, and the only possible solution of the simultaneous equations is

$$\alpha_1 = \alpha_2 = \alpha_3 = 0$$

Hence, the vectors $\mathbf{A}_1$, $\mathbf{A}_2$, and $\mathbf{A}_3$ are linearly independent.

Since $\mathbf{A}_1$, $\mathbf{A}_2$, and $\mathbf{A}_3$ are linearly independent, any subset of them is also linearly independent. For example, consider the vectors $\mathbf{A}_1$ and $\mathbf{A}_2$. If these vectors were linearly dependent, the following simultaneous equations would have nontrivial solutions:

$$2\alpha_1 - 3\alpha_2 = 0$$
$$-\alpha_1 + \alpha_2 = 0$$
$$3\alpha_1 + 2\alpha_2 = 0$$

However, it is not difficult to verify that the only solution of these equations is

$$\alpha_1 = \alpha_2 = 0$$

which means that $\mathbf{A}_1$ and $\mathbf{A}_2$ are linearly independent. The same conclusion applies to the subsets $\mathbf{A}_1$ and $\mathbf{A}_3$, and $\mathbf{A}_2$ and $\mathbf{A}_3$.

Example 2. Are the row vectors $\mathbf{B}_1$, $\mathbf{B}_2$, **and** $\mathbf{B}_3$ linearly dependent or independent?

$$\mathbf{B}_1 = [4 \quad 1 \quad -3] \qquad \mathbf{B}_2 = [3 \quad -8 \quad 9] \qquad \mathbf{B}_3 = [1 \quad 2 \quad -3]$$

When these vectors are substituted into Equation (5-13), the following simultaneous equations are obtained:

$$4\alpha_1 + 3\alpha_2 + \alpha_3 = 0$$
$$\alpha_1 - 8\alpha_2 + 2\alpha_3 = 0$$
$$-3\alpha_1 + 9\alpha_2 - 3\alpha_3 = 0$$

These equations have nontrivial solutions because the coefficient matrix is singular; the solutions are as follows:

$$\begin{bmatrix} \alpha_1 \\ \alpha_2 \\ \alpha_3 \end{bmatrix} = \beta \begin{bmatrix} -2 \\ 1 \\ 5 \end{bmatrix}$$

Therefore, the relationship among the vectors is

$$-2\mathbf{B}_1 + \mathbf{B}_2 + 5\mathbf{B}_3 = \mathbf{0}$$

and the three vectors are linearly dependent.

Example 3. Express the vector $\mathbf{B}_3$ in the preceding example as a linear combination of $\mathbf{B}_1$ and $\mathbf{B}_2$.

The following relation is obtained from Equation (5-15) and the results of Example 2:

$$\mathbf{B}_3 = \frac{2}{5}\mathbf{B}_1 - \frac{1}{5}\mathbf{B}_2$$

Thus, $\mathbf{B}_3$ has been expressed as a linear combination of the other vectors. In an analogous manner, $\mathbf{B}_1$ and $\mathbf{B}_2$ can be written as linear combinations of the remaining two vectors.

5.7. Rank of a Matrix. As discussed previously in Section 3.2, minor determinants can be formed from the determinant of a square matrix by deleting an equal number of rows and columns from the original determinant. Such determinants may have any order from one up to the order of the square matrix itself. If a matrix is rectangular instead of square, it is also possible to form determinants from its elements. The largest determinants* that can be formed will have the same order as the number of rows or the number of columns, whichever is smaller. Other determinants can be formed by deleting a sufficient number of rows and columns to produce square determinants. For instance, consider the following 3 X 5 matrix:

$$\mathbf{A} = \begin{bmatrix} A_{11} & A_{12} & A_{13} & A_{14} & A_{15} \\ A_{21} & A_{22} & A_{23} & A_{24} & A_{25} \\ A_{31} & A_{32} & A_{33} & A_{34} & A_{35} \end{bmatrix}$$

The largest determinants that can be formed are of third order. They can be obtained by deleting any two columns from the matrix, for instance, columns 1 and 2, columns 1 and 3, columns 1 and 4, etc., up to columns 4 and 5. In this manner, a total of ten third-order determinants can be obtained. Second-order determinants can be formed from the elements of the matrix **A** by deleting any one row and any three columns; a total of 30 second-order determinants can be formed in this way. Finally, the first-order determinants are obtained by deleting any two rows and any four columns. The number of first-order determinants is, of course, the same as the number of elements in the matrix.

The *rank* of a matrix is defined as the order of the largest nonzero determinant that can be obtained from the elements of the matrix. This definition applies to both rectangular and square matrices. The rank of a matrix can be found by beginning with the largest order determinants and evaluating each of them to ascertain whether any of them is nonzero. Assume that the order of the largest determinants is n. If a nonzero nth order determinant is found, then the matrix is said to have rank n. If not, then the process is repeated for all determinants of order $n - 1$. If a nonzero determinant is found among these determinants, then the matrix has rank $n - 1$. If not, the procedure can be repeated for successively lower order determinants until a nonzero determinant is found, and the rank is thereby determined. The rank of a matrix **A** is written as $r(\mathbf{A})$.

*In this section the expression "largest determinant" refers to the order of the determinant, not its numerical value.

As an illustration, consider the following third-order square matrix **A** and assume that its rank is to be determined:

$$\mathbf{A} = \begin{bmatrix} 5 & 2 & -3 \\ -10 & 4 & 7 \\ -15 & -6 & 9 \end{bmatrix}$$

The largest determinant that can be formed from this matrix is of order three, and it has the value zero. Therefore, the rank of the matrix must be two or less. There are nine second-order determinants that can be obtained from the elements of the matrix **A**; specifically, these determinants are the minors that correspond to the nine elements of the matrix. While some of these determinants are zero, it can be noted immediately that there are several having nonzero values. For instance, the determinant that remains when the first row and first column are deleted has the value 78. Thus, the matrix **A** has rank two, which is written as follows:

$$r(\mathbf{A}) = 2$$

If all nine second-order determinants were equal to zero, then the rank would be one, because the determinant obviously has at least one nonzero element. Of course, if all elements in a matrix are zero, its rank is zero.

An example of a rectangular matrix is the following:

$$\mathbf{B} = \begin{bmatrix} 2 & 4 & -1 & 7 \\ 3 & 6 & -5 & 7 \\ -4 & 3 & 2 & 0 \end{bmatrix}$$

There are four third-order determinants in this matrix; moreover, at least one of these determinants is nonzero (for example, the determinant obtained by deleting the first column). Thus, the matrix **B** has rank three:

$$r(\mathbf{B}) = 3$$

If all four determinants of order three were equal to zero, then the determinants of order two would be investigated, and so forth.

In Chapter 4, a singular matrix was defined as a square matrix having a determinant equal to zero (see Equation 4–10). From the definition of the rank of a matrix, it is apparent that a square matrix of order n has a rank less than n if the matrix is singular. Conversely, a square matrix is singular if its rank is less than n and nonsingular if its rank equals n.

The rank of a matrix is related to the linear dependence or independence of the rows of the matrix. It should be recalled from Chapter 3 that the value of a determinant is zero if one row is a linear combination of other rows. If this condition exists, then the several rows involved in the linear combination are linearly dependent since (with each row considered as a vector) they satisfy Equation (5-13). It follows also that if some of the rows are linearly dependent, then all of the rows are linearly dependent, for the reasons explained in the preceding section. Hence, the conclusion is reached that the value of a determinant is zero if its rows are linearly dependent. Of course, the same conclusions apply to the columns of a determinant. Thus, a square matrix whose rows (or columns) are linearly dependent will be singular.

An important relationship to be noted is that the maximum number of linearly independent columns in a matrix is the same as the maximum number of linearly independent rows. To show this fact, assume that a matrix **A** (either rectangular or square) has exactly k linearly independent rows. Then all minor determinants formed from the elements of **A** and having order $k + 1$ or greater must be equal to zero, while there is at least one determinant of order k that does not vanish. Now consider the transpose of the matrix **A**, and recall that the value of a determinant is not changed by transposition (see Equation 3-5). In the transposed matrix $\mathbf{A}^\mathsf{T}$, there must be at least one determinant of order k that does not vanish, while all determinants of order $k + 1$ or greater must vanish. Thus, the maximum number of linearly independent rows in $\mathbf{A}^\mathsf{T}$ is k. But, since the number of rows of $\mathbf{A}^\mathsf{T}$ is the same as the number of columns of **A**, it follows that the maximum number of linearly independent columns of the original matrix is the same as the number of linearly independent rows.

Finally, it can be seen that the rank of a matrix is the same as the maximum number of linearly independent rows or columns. This relationship arises from the definition of the rank of a matrix (that is, the rank is the order of the largest nonvanishing determinant). Since a nonvanishing determinant requires that the rows and columns be linearly independent, the rank of a matrix is determined by the maximum number of such linearly independent rows and columns in the matrix.

The concepts of linear dependence and the rank of a matrix will be used in the next section to discuss the nature of the solutions of simultaneous equations.

5.8. Conditions for the Solution of Equations. The solutions of simultaneous equations (both homogeneous equations and nonhomogeneous equations) can take various forms, depending upon the nature of the equations. The form of the solution, or even whether or not there is a solution, can be determined by investigating the rank of two matrices. The first matrix is the coefficient matrix **A** appearing in Equations (5-2) and (5-10). The second matrix, called the *augmented matrix*, is obtained by augmenting the matrix **A** by one column consisting of the vector **B** of right-hand sides. Thus, for n equations in n unknowns (see Equations 5-1 and 5-2) the augmented matrix, which is denoted $\mathbf{A}^b$, is

$$\mathbf{A}^b = \begin{bmatrix} A_{11} & A_{12} & \dots & A_{1n} & B_1 \\ A_{21} & A_{22} & \dots & A_{2n} & B_2 \\ \dots & \dots & \dots & \dots & \dots \\ A_{n1} & A_{n2} & \dots & A_{nn} & B_n \end{bmatrix} \tag{5-16}$$

and has n rows and $n + 1$ columns.

If the coefficient matrix $\mathbf{A}$ is nonsingular, which means that its rank is the same as its order n, then the rank of the augmented matrix $\mathbf{A}^b$ also is n; that is,

$$\text{If } r(\mathbf{A}) = n, \text{ then } r(\mathbf{A}^b) = n \tag{5-17}$$

This conclusion holds for both homogeneous ($\mathbf{B} = \mathbf{0}$) and nonhomogeneous ($\mathbf{B} \neq \mathbf{0}$) equations, and it follows from the fact that the augmented matrix contains the matrix of coefficients. For the same reason, it is not possible for the rank of $\mathbf{A}$ to exceed the rank of $\mathbf{A}^b$:

$$r(\mathbf{A}) \leqslant r(\mathbf{A}^b) \tag{5-18}$$

Finally, the rank of $\mathbf{A}^b$ can never exceed the rank of $\mathbf{A}$ by more than one; hence,

$$\text{If } r(\mathbf{A}^b) > r(\mathbf{A}), \text{ then } r(\mathbf{A}^b) = r(\mathbf{A}) + 1 \tag{5-19}$$

The validity of this last statement can be seen from Equation (5-16) by observing that the "B" column can supply no more than one linearly independent column, and hence it can raise the rank by no more than one. The preceding three relationships between $r(\mathbf{A})$ and $r(\mathbf{A}^b)$ will be evident in the ensuing discussions of the solutions of simultaneous equations.

A set of simultaneous equations is said to be *consistent* if there is a solution to the equations and *inconsistent* if there is no solution. Of course, in this context the trivial solution is considered to be a valid solution. A set of n simultaneous equations in n unknowns (see Equations 5-2 and 5-10) is consistent and has a solution if the rank $r(\mathbf{A})$ of the coefficient matrix and the rank $r(\mathbf{A}^b)$ of the augmented matrix are the same. If the rank of $\mathbf{A}$ is less than the rank of $\mathbf{A}^b$, that is, if $r(\mathbf{A}) = r(\mathbf{A}^b) - 1$, then the equations are inconsistent and have no solution.

If the equations are consistent, that is, if $r(\mathbf{A}) = r(\mathbf{A}^b)$, then there is only one rank to be considered, and it may be called the *rank of the equations*. If this rank equals n, which means that the matrix $\mathbf{A}$ is nonsingular, then there is a unique solution to the equations. However, if the rank of the equations is r, where r is less than n, then $\mathbf{A}$ is singular, and there exists an infinite number of solutions to the equations. To be specific, there will be $n - r$ unknowns that can be assigned arbitrary values, and the remaining r unknowns will be related uniquely to them. In other words, if the solution of the equations is expressed as a vector $\mathbf{X}$ (see Equation 5-2 or 5-10), then $\mathbf{X}$ will be a function of $n - r$ arbitrary constants.

All of the preceding rules apply to both homogeneous and nonhomogeneous equations and will be illustrated in the examples to follow. It is important to note, however, that the ranks of the coefficient matrix and the augmented matrix are always the same when the equations are homogeneous; therefore, homogeneous equations always are consistent and have at least one solution (the trivial solution). For convenient reference, the conditions for the solution of simultaneous equations are summarized in Table 5-1.

If the set of equations has rank r less than n, the solution is of a more complicated nature. Consider first the case of homogeneous equations (with $\mathbf{B} = \mathbf{0}$). The *complete solution* of such equations must contain $n - r$ arbitrary constants, and it always can be expressed as a linear combination of $n - r$ linearly independent solution vectors. For instance, the equations given in Example 2 of Section 5.5 have $n = 2$ and $r = 1$; thus, $n - r = 1$ and the solution is expressed as an arbitrary constant β times one independent solution vector of Equation (5-10). Similarly, the equations of Example 3 have $n = 3$ and $r = 2$, so that again $n - r = 1$ and the solution contains one arbitrary constant. An example will be given later of homogeneous equations having $n = 3$ and $r = 1$, so that $n - r = 2$ and the solution consists of two linearly independent solution vectors, each multiplied by an arbitrary constant. If one of those constants is taken as zero, what remains is still a solution of the homogeneous equations but it is only a *partial solution*, not the complete solution. The complete solution must always contain $n - r$ linearly independent solution vectors with a corre-

Table 5-1

Conditions for Solution of n Simultaneous Equations

Type of Equations	Conditions of Rank	Nature of Solution
Consistent equations	$r(\mathbf{A}) = r(\mathbf{A}^b) = n$	Unique solution
	$r(\mathbf{A}) = r(\mathbf{A}^b) = r < n$	Infinite number of solutions with $n - r$ arbitrary constants
Inconsistent equations	$r(\mathbf{A}) < r(\mathbf{A}^b)$	No solution
Homogeneous equations (special case of consistent equations)	$r(\mathbf{A}) = r(\mathbf{A}^b) = n$	Unique solution (the trivial solution)
	$r(\mathbf{A}) = r(\mathbf{A}^b) = r < n$	Infinite number of solutions with $n - r$ arbitrary constants

sponding number of arbitrary constants. Every possible solution to the equations is contained in the complete solution.

The choice of the $n - r$ linearly independent solution vectors is not unique, because any set of linearly independent vectors can be replaced by an equivalent but different set such that each of the new vectors is a linear combination of the original ones. Since this replacement can be done in an infinite number of ways, there is an infinite number of different sets of $n - r$ linearly independent solution vectors (with each vector multiplied by an arbitrary constant).

Now consider a set of nonhomogeneous equations (with $\mathbf{B} \neq \mathbf{0}$). In this case, it is convenient to separate the solution of the equations into two parts, a *homogeneous solution* $\mathbf{X}_\mathrm{H}$ and a *particular solution* $\mathbf{X}_\mathrm{P}$. Then the *general solution* $\mathbf{X}$ is the sum of the homogeneous and particular solutions:

$$\mathbf{X} = \mathbf{X}_\mathrm{H} + \mathbf{X}_\mathrm{P} \tag{5-20}$$

The homogeneous solution is the complete solution of the corresponding homogeneous equations, and the particular solution is any solution that produces the vector $\mathbf{B}$ on the right-hand side of Equation (5-2). Thus, $\mathbf{X}_\mathrm{H}$ satisfies the equation

$$\mathbf{A}\mathbf{X}_\mathrm{H} = \mathbf{0} \tag{5-21}$$

and $\mathbf{X}_\mathrm{P}$ satisfies the equation

$$\mathbf{A}\mathbf{X}_\mathrm{P} = \mathbf{B} \tag{5-22}$$

Summation of these two equations yields

$$\mathbf{A}(\mathbf{X}_\mathrm{H} + \mathbf{X}_\mathrm{P}) = \mathbf{0} + \mathbf{B} = \mathbf{B}$$

which agrees with the original matrix equation $\mathbf{AX} = \mathbf{B}$.

Because the homogeneous solution $\mathbf{X}_\mathrm{H}$ must be the complete solution of Equation (5-21), it must contain $n - r$ linearly independent solution vectors, each multiplied by an arbitrary constant. As previously explained, the choice of these vectors is not unique. The vector $\mathbf{X}_\mathrm{P}$ is any particular solution that satisfies Equation (5-22). It is not unique either, because there is an infinite number of particular solutions satisfying Equation (5-22) when r is less than n. The general solution $\mathbf{X}$ contains all possible solutions to the equation $\mathbf{AX} = \mathbf{B}$, no matter what particular solution $\mathbf{X}_\mathrm{P}$ is chosen and no matter which set of linearly independent solution vectors is used in the homogeneous solution $\mathbf{X}_\mathrm{H}$.

The preceding concepts will now be illustrated with examples of two and three simultaneous equations. Consider first the following two equations:

$$-X_1 + 4X_2 = 8$$

$$X_1 + X_2 = 7$$

In this example, the coefficient matrix $\mathbf{A}$ and the augmented matrix $\mathbf{A}^b$ are

$$\mathbf{A} = \begin{bmatrix} -1 & 4 \\ 1 & 1 \end{bmatrix} \qquad \mathbf{A}^b = \begin{bmatrix} -1 & 4 & 8 \\ 1 & 1 & 7 \end{bmatrix}$$

Each of these matrices has rank equal to two; hence, the equations are consistent and have a unique solution. The solution is found to be $X_1 = 4$ and $X_2 = 3$, or, in vector form,

$$\mathbf{X} = \begin{bmatrix} 4 \\ 3 \end{bmatrix} \tag{a}$$

The two equations can be interpreted geometrically as the equations of two intersecting straight lines (see Figure 5-1), and their point of intersection represents the solution of the equations.

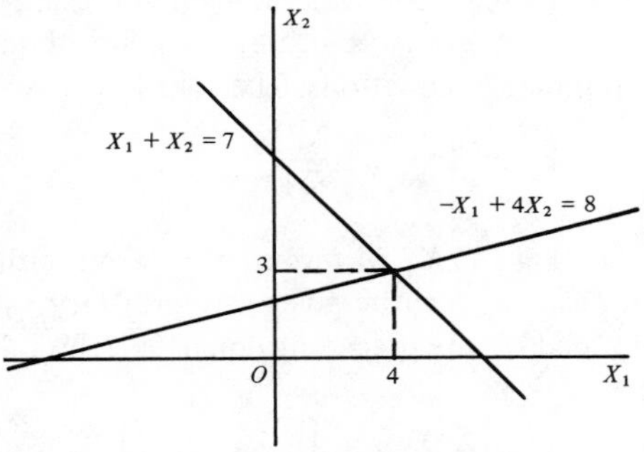

Figure 5-1. Graph of two simultaneous equations having $r(\mathbf{A}) = r(\mathbf{A}^b) = 2$

Next, consider the following two equations:

$$X_1 - 3X_2 = -3$$

$$X_1 - 3X_2 = -18$$

Since the rank of the coefficient matrix is one and the rank of the augmented matrix is two, the equations are inconsistent and have no solution. The equations represent the two parallel lines shown in Figure 5-2.

Another possibility for two nonhomogeneous simultaneous equations is that both the coefficient matrix and the augmented matrix have the same rank, but this rank is equal to one, which is less than the number of equations. These conditions are encountered with the two equations shown:

$$3X_1 + 2X_2 = 12$$

$$-6X_1 - 4X_2 = -24$$

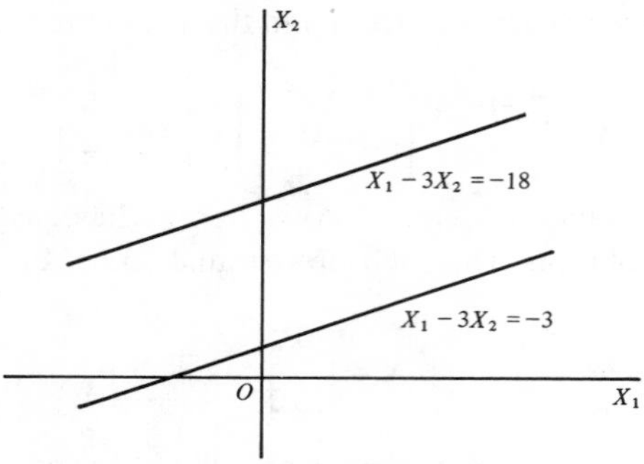

Figure 5-2. Graph of two simultaneous equations having $r(\mathbf{A}) = 1$ and $r(\mathbf{A}^b) = 2$

Since the ranks of $\mathbf{A}$ and $\mathbf{A}^b$ are the same, the equations are consistent and must have a solution. Moreover, there is an infinite number of solutions because the rank is less than the number of equations. The solution is

$$X_1 = -\frac{2}{3}X_2 + 4$$

as can be found by solving for X_1 in terms of X_2 from either of the two equations. Thus, the unknown X_2 can be given any arbitrary value, whereupon the value of X_1 is determined by the preceding equation. The solution vector $\mathbf{X}$ may be stated as follows:

$$\mathbf{X} = \begin{bmatrix} X_1 \\ X_2 \end{bmatrix} = \begin{bmatrix} -\frac{2}{3}X_2 + 4 \\ X_2 \end{bmatrix}$$

$$= X_2 \begin{bmatrix} -\frac{2}{3} \\ 1 \end{bmatrix} + \begin{bmatrix} 4 \\ 0 \end{bmatrix}$$

or, substituting an arbitrary constant β for $-X_2/3$,

$$\mathbf{X} = \beta \begin{bmatrix} 2 \\ -3 \end{bmatrix} + \begin{bmatrix} 4 \\ 0 \end{bmatrix} \tag{b}$$

in which β can have any value. This result can be verified easily by premultiplying $\mathbf{X}$ by the matrix $\mathbf{A}$ of coefficients. The two simultaneous equations represent the same straight line on the graph (see Figure 5-3), and hence all points on that line represent solutions of the equations. For instance, if $\beta = -1$ the solution is

$$\mathbf{X} = \begin{bmatrix} 2 \\ 3 \end{bmatrix}$$

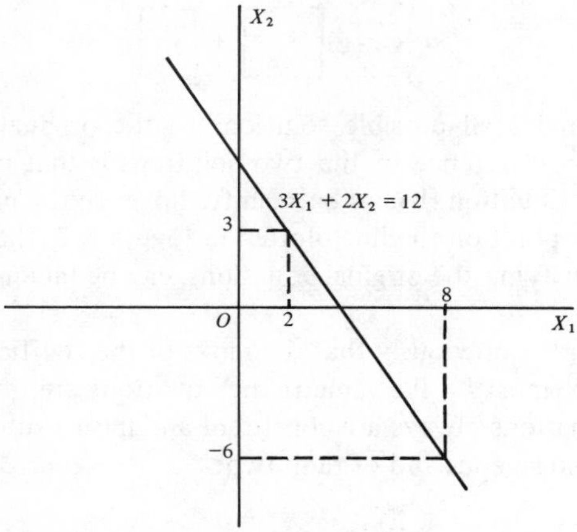

Figure 5-3. Graph of two simultaneous equations having $r(\mathbf{A}) = r(\mathbf{A}^b) = 1$

and if $\beta = 2$ the solution is

$$\mathbf{X} = \begin{bmatrix} 8 \\ -6 \end{bmatrix}$$

as shown in Figure 5-3.

In the preceding example, the rank of the equations is $r = 1$, hence $n - r = 1$. Note that the number of arbitrary constants in the general solution $\mathbf{X}$ (see Equation b) equals $n - r$. The homogeneous solution is

$$\mathbf{X}_\mathrm{H} = \beta \begin{bmatrix} 2 \\ -3 \end{bmatrix}$$

which is the complete solution (consisting of one linearly independent solution vector) of the corresponding homogeneous equations. A particular solution is

$$\mathbf{X}_\mathrm{P} = \begin{bmatrix} 4 \\ 0 \end{bmatrix}$$

as shown in Equation (b). However, there is an infinite number of possible particular solutions, any one of which may be used in the general solution. For example, another particular solution is

$$\mathbf{X}_\mathrm{P} = \begin{bmatrix} 2 \\ 3 \end{bmatrix}$$

which gives the following general solution:

$$\mathbf{X} = \beta \begin{bmatrix} 2 \\ -3 \end{bmatrix} + \begin{bmatrix} 2 \\ 3 \end{bmatrix} \qquad \text{(c)}$$

This solution contains all possible solutions to the original equations, as does Equation (b). The difference in the two solutions is that part of the homogeneous solution in Equation (b) has been shifted over to the particular solution in Equation (c). Any point on the line plotted in Figure 5-3, that is, any two values of X_1 and X_2 satisfying the original equations, can be taken as a particular solution $\mathbf{X}_\text{P}$.

It was observed previously that the ranks of the coefficient matrix and the augmented matrix must be the same if the equations are homogeneous; hence, homogeneous equations always are consistent and have a solution. The next two equations are homogeneous and of rank two:

$$2X_1 - 5X_2 = 0$$

$$X_1 - X_2 = 0$$

Because the rank is the same as the number of equations, these equations can have only one solution; namely, the trivial solution. The equations are plotted in Figure 5-4, and it is seen that the lines intersect at the origin.

The following homogeneous equations are of rank one:

$$X_1 - 2X_2 = 0$$

$$-3X_1 + 6X_2 = 0$$

Since the rank is less than the number of equations, the coefficient matrix is singular and there is an infinite number of solutions to the equations. The solution takes the form

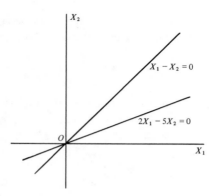

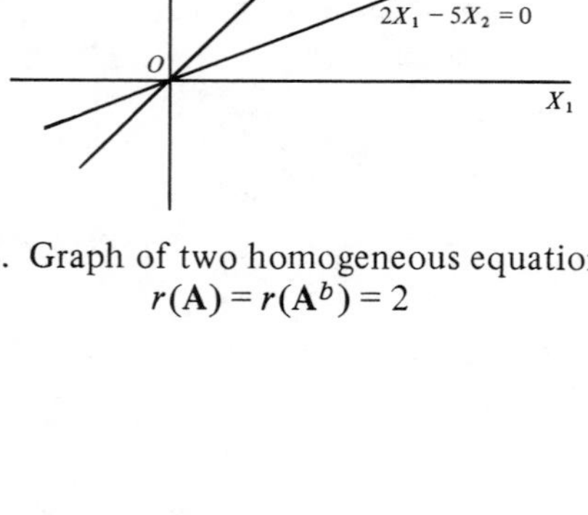

Figure 5-4. Graph of two homogeneous equations having $r(\mathbf{A}) = r(\mathbf{A}^b) = 2$

$$X_1 = 2X_2$$

or

$$\mathbf{X} = \beta \begin{bmatrix} 2 \\ 1 \end{bmatrix} \tag{d}$$

in which β is arbitrary. Thus, the ratio of X_1 to X_2 is always the same, but the magnitude of the vector is indefinite. The two simultaneous equations represent the same straight line through the origin (see Figure 5-5), and any point on that line represents a solution. Since $n - r = 1$, there is one arbitrary constant in the complete solution (Equation d).

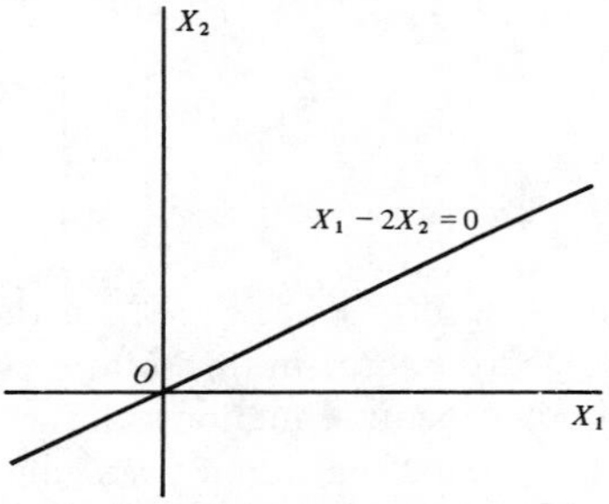

Figure 5-5. Graph of two homogeneous equations having $r(\mathbf{A}) = r(\mathbf{A}^b) = 1$

Now suppose that a set of three simultaneous equations has rank 3, that is, $r(\mathbf{A}) = r(\mathbf{A}^b) = 3$. (All of the sets of three simultaneous equations given in the text and problems for Sections 5.2 through 5.4 satisfy this particular condition.) Then it follows immediately that the equations are consistent and have a unique solution. A geometrical interpretation for such a set of equations is readily available, because each equation represents a plane in three-dimensional space. Each pair of planes intersects in a straight line, and all three lines intersect at a common point; this point is the intersection point of the three planes and represents the unique solution of the three equations.

If the rank of three simultaneous equations is $r(\mathbf{A}) = r(\mathbf{A}^b) = 2$, then all three planes intersect in the same straight line. The three planes may be distinct from one another or two of them may coincide. There is an infinite number of solutions to the equations, and these are represented by all points along the line of intersection. Because $n = 3$ and $r = 2$, the solution will have one arbitrary constant (see Table 5-1). The following equations illustrate this situation:

$$X_1 - 3X_2 + 5X_3 = 1$$

$$-2X_1 + 7X_2 - 2X_3 = -8$$

$$4X_1 - 13X_2 + 12X_3 = 10$$

These equations have rank two for both the coefficient matrix **A** and the augmented matrix $\mathbf{A}^b$; their solution can be found by elimination to be:

$$X_1 = -29X_3 - 17 \qquad X_2 = -8X_3 - 6 \qquad X_3 = X_3$$

Thus, X_3 may be treated as an arbitrary unknown constant, while X_1 and X_2 are related to it. The solution can be expressed in the vector form

$$\mathbf{X} = \begin{bmatrix} X_1 \\ X_2 \\ X_3 \end{bmatrix} = \begin{bmatrix} -29X_3 - 17 \\ -8X_3 - 6 \\ X_3 \end{bmatrix}$$

$$= \beta \begin{bmatrix} 29 \\ 8 \\ -1 \end{bmatrix} + \begin{bmatrix} -17 \\ -6 \\ 0 \end{bmatrix} \tag{e}$$

in which β has been introduced in place of $-X_3$ as the arbitrary constant. As explained previously, the first vector in this solution is the complete solution of the corresponding homogeneous equations, the second is a particular solution, and their sum is the general solution. By assigning values to the constant β, various particular solutions of the equations can be obtained. Any of these particular solutions may be used with the homogeneous solution to give a general solution. For instance, the following general solution

$$\mathbf{X} = \beta \begin{bmatrix} 29 \\ 8 \\ -1 \end{bmatrix} + \begin{bmatrix} 12 \\ 2 \\ -1 \end{bmatrix} \tag{f}$$

contains a particular solution obtained by setting $\beta = 1$ in Equation (e). Both Equations (e) and (f) contain all possible solutions to the three simultaneous equations.

Now assume that the three equations have ranks such that $r(\mathbf{A}) = r(\mathbf{A}^b) = 1$. In that event, all three planes represented by the equations coincide, and again there is an infinite number of solutions (represented by the plane itself). The solution will have two arbitrary unknowns, as in the following example:

$$\begin{aligned} X_1 - 2X_2 + 3X_3 &= 4 \\ 4X_1 - 8X_2 + 12X_3 &= 16 \\ -2X_1 + 4X_2 - 6X_3 &= -8 \end{aligned}$$

The solution is

$$X_1 = 2X_2 - 3X_3 + 4$$

with X_2 and X_3 arbitrary. In vector form the solution can be written

$$\mathbf{X} = \begin{bmatrix} X_1 \\ X_2 \\ X_3 \end{bmatrix} = \begin{bmatrix} 2X_2 - 3X_3 + 4 \\ X_2 \\ X_3 \end{bmatrix} \tag{g}$$

$$= \beta_1 \begin{bmatrix} 2 \\ 1 \\ 0 \end{bmatrix} + \beta_2 \begin{bmatrix} -3 \\ 0 \\ 1 \end{bmatrix} + \begin{bmatrix} 4 \\ 0 \\ 0 \end{bmatrix} \tag{h}$$

in which β_1 and β_2 have been introduced as arbitrary constants in place of X_2 and X_3. The first two vectors constitute the homogeneous solution $\mathbf{X}_\mathrm{H}$, which is the complete solution of the corresponding homogeneous equations and contains $n - r = 2$ arbitrary constants associated with $n - r$ linearly independent solution vectors. The last vector is a particular solution $\mathbf{X}_\mathrm{P}$. Other particular solutions can be obtained by assigning numerical values to β_1 and β_2 in Equation (h), and any such vectors can be taken as a particular solution in the general solution (Equation h) if desired.

The choice of the two vectors constituting the homogeneous solution (see Equation h) is not unique. Any two linearly independent vectors satisfying Equation (g) can be used. For instance, another possibility for the general solution is

$$\mathbf{X} = \beta_3 \begin{bmatrix} -1 \\ 1 \\ 1 \end{bmatrix} + \beta_4 \begin{bmatrix} -5 \\ -1 \\ 1 \end{bmatrix} + \begin{bmatrix} 4 \\ 0 \\ 0 \end{bmatrix} \tag{i}$$

in which β_3 and β_4 can be given any values. This solution for $\mathbf{X}$ contains all possible solutions to the original equations, as also does Equation (h). Of course, each of the first two vectors in Equation (i) is a linear combination of the corresponding vectors in Equation (h).

Continuing now with further examples of three equations, suppose that $r(\mathbf{A}) = 2$ and $r(\mathbf{A}^b) = 3$, as in the following case:

$$7X_1 + 2X_2 - 3X_3 = 5$$

$$X_1 - 4X_2 + X_3 = 2$$

$$6X_1 + 6X_2 - 4X_3 = 9$$

These equations are inconsistent, and no solution can be found. In geometrical terms, two of the equations represent planes that intersect in a line, whereas the third one represents a plane that is parallel to the line but does not pass through the line. Thus, there is no point that is common to all three planes.

Another case of inconsistency occurs when $r(\mathbf{A}) = 1$ and $r(\mathbf{A}^b) = 2$; in this event, all three planes are parallel. The three parallel planes may be distinct or two of them may coincide. It is not possible to have the case where $r(\mathbf{A}) = 1$ and $r(\mathbf{A}^b) = 3$, as shown by Equation (5-19).

If the three equations are homogeneous and have rank equal to three, there will be one solution (the trivial solution). The equations represent three different planes, each of which passes through the origin (which is the only point common to all three planes). If the rank is equal to two, there is one arbitrary constant in the solution (see Example 3 of Section 5.5). In this case, all three planes intersect in a line through the origin; the three planes may be distinct or two of them may coincide. If the rank is one, there are two arbitrary constants in the solution; then all three planes coincide and pass through the origin.

An illustration of the last case is the following:

$$-2X_1 + X_2 + 2X_3 = 0$$
$$4X_1 - 2X_2 - 4X_3 = 0$$
$$-10X_1 + 5X_2 + 10X_3 = 0$$

The solution of these equations is

$$X_1 = \frac{1}{2}X_2 + X_3$$

or

$$\mathbf{X} = \beta_1 \begin{bmatrix} 1 \\ 2 \\ 0 \end{bmatrix} + \beta_2 \begin{bmatrix} 1 \\ 0 \\ 1 \end{bmatrix}$$

in which β_1 and β_2 can be assigned any desired values. If β_2 is set equal to zero, a solution remains that produces an infinite number of solutions (because β_1 can be assigned any arbitrary value). However, it is not the complete solution and cannot produce all possible solutions of the equations. In geometric terms, the complete solution includes all points on the common plane, whereas the solution with β_2 equal to zero is only a line in that plane.

Sometimes a situation arises in which there are more unknowns than equations; for instance, there might be three unknowns but only two simultaneous equations. These equations can be cast into the standard form of three equations in three unknowns by adding a third equation with all coefficients and the right-hand side equal to zero. The rank $r(\mathbf{A})$ of the coefficient matrix for this new set of equations is clearly less than the number of equations, hence there is an infinite number of solutions (unless the equations happen to be inconsistent).

The concepts illustrated in the preceding examples for two and three simultaneous equations can be applied to any number of equations. The existence and nature of the solutions can be determined by reference to Table 5-1. In most cases that arise in engineering work, the equations have a unique solution if they are nonhomogeneous, and they have rank equal to $n - 1$ if they are homogeneous.

Problems

5.2-1. Solve the following simultaneous equations by inversion:

$$3X_1 + 5X_2 = 5$$
$$2X_1 + 4X_2 = 2$$

5.2-2. Solve the following simultaneous equations by inversion:

$$2X_1 - X_2 = -10$$
$$2X_1 + 3X_2 = 6$$

5.2-3. Solve the following simultaneous equations by inversion:

$$6X_1 + 13X_2 = 10$$
$$-3X_1 - 7X_2 = -7$$

5.2-4. Solve the following simultaneous equations by inversion:

$$X_1 - 2X_2 - X_3 = 10$$
$$2X_1 + 3X_2 - 2X_3 = -1$$
$$-4X_1 + X_2 - 5X_3 = -1$$

5.2-5. Solve the following simultaneous equations by inversion:

$$4X_1 + 2X_2 - X_3 = -3$$
$$2X_1 + 3X_2 + 2X_3 = 11$$
$$X_1 - 5X_2 - 2X_3 = 11$$

5.2-6. Solve the following simultaneous equations by inversion:

$$2X_1 - 4X_2 + 5X_3 = 5$$
$$X_1 + 3X_2 - 2X_3 = -10$$
$$3X_1 + 2X_2 + 4X_3 = 6$$

5.3-1. Solve the simultaneous equations given in Problem 5.2-1 by Cramer's rule.

5.3-2. Solve the simultaneous equations given in Problem 5.2-2 by Cramer's rule.

5.3-3. Solve the simultaneous equations given in Problem 5.2-3 by Cramer's rule.

5.3–4. Solve the simultaneous equations given in Problem 5.2–4 by Cramer's rule.

5.3–5. Solve the simultaneous equations given in Problem 5.2–5 by Cramer's rule.

5.3–6. Solve the simultaneous equations given in Problem 5.2–6 by Cramer's rule.

5.4–1. Solve the simultaneous equations given in Problem 5.2–4 by the method of elimination.

5.4–2. Solve the simultaneous equations given in Problem 5.2–5 by the method of elimination.

5.4–3. Solve the simultaneous equations given in Problem 5.2–6 by the method of elimination.

5.4–4. Solve the following simultaneous equations by the method of elimination:

$$5X_1 - 2X_2 - X_3 = 19$$
$$X_1 + 3X_2 + 3X_3 = 12$$
$$8X_1 + 4X_2 + X_3 = 6$$

5.4–5. Solve the following simultaneous equations by the method of elimination:

$$27.2X_1 - 9.2X_2 + 7.1X_3 = 117.6$$
$$6.8X_1 + 16.8X_2 + 12.4X_3 = 25.4$$
$$-3.2X_1 + 14.1X_2 + 21.5X_3 = 30.0$$

5.4–6. Solve the following simultaneous equations by the method of elimination:

$$247X_1 + 87X_2 + 54X_3 = 2624$$
$$72X_1 - 129X_2 + 103X_3 = 8098$$
$$-43X_1 + 80X_2 + 322X_3 = 7156$$

5.4–7. Solve the following simultaneous equations by the method of elimination:

$$0.607X_1 + 0.982X_2 - 0.382X_3 = -1.121$$
$$0.318X_1 - 0.880X_2 + 0.165X_3 = 2.548$$
$$0.372X_1 - 0.527X_2 - 0.749X_3 = 1.034$$

5.4–8. Solve the following simultaneous equations by the method of elimination:

$$4X_1 + X_2 = 18$$
$$X_1 + 2X_2 + X_3 = 4$$
$$X_2 + 2X_3 + X_4 = 11$$
$$X_3 + 4X_4 = 31$$

5.4-9. Solve the following simultaneous equations by the method of elimination:

$$
\begin{aligned}
2X_1 + X_2 &= -1 \\
X_1 - 3X_2 + 2X_3 &= -22 \\
X_2 - 3X_3 + 2X_4 &= 31 \\
X_3 + 4X_4 &= 38
\end{aligned}
$$

5.4-10. Solve the following simultaneous equations by the method of elimination:

$$
\begin{aligned}
3X_1 + 7X_2 - 2X_3 + 3X_4 &= 4 \\
-X_1 - 2X_2 + 4X_3 - 3X_4 &= 5 \\
2X_1 - 5X_2 - 2X_3 + 2X_4 &= 9 \\
-6X_1 + 3X_2 - X_3 + 2X_4 &= -10
\end{aligned}
$$

5.4-11. Solve the following simultaneous equations by the method of elimination:

$$
\begin{aligned}
2.17X_1 + 1.16X_2 - 0.92X_3 + 1.33X_4 &= 13.88 \\
-1.73X_1 + 4.91X_2 + 2.09X_3 + 2.56X_4 &= 0 \\
2.22X_1 + 1.45X_2 - 3.74X_3 + 3.41X_4 &= 26.55 \\
1.10X_1 - 2.28X_2 + 1.83X_3 + 1.29X_4 &= -2.59
\end{aligned}
$$

5.4-12. Solve the following simultaneous equations by the method of elimination:

$$
\begin{aligned}
-3X_1 + 5X_2 &= 21 \\
2X_1 + 6X_2 - 5X_3 &= 9 \\
X_2 + 4X_3 + 3X_4 &= -5 \\
-6X_3 + 2X_4 + 3X_5 &= 1 \\
-X_4 + 3X_5 &= 19
\end{aligned}
$$

5.4-13. Solve the following simultaneous equations by the method of elimination:

$$
\begin{aligned}
2X_1 + X_2 &= 4 \\
X_1 + 4X_2 + X_3 &= -6 \\
X_2 + 4X_3 + X_4 &= 14 \\
X_3 + 4X_4 + X_5 &= -10 \\
X_4 + 4X_5 + X_6 &= 18 \\
X_5 + 2X_6 &= -8
\end{aligned}
$$

5.5-1. Obtain the solution of the following homogeneous equations:

$$8X_1 + 2X_2 - X_3 = 0$$
$$-3X_1 - 4X_2 + 2X_3 = 0$$
$$2X_1 - 3X_2 + 5X_3 = 0$$

5.5-2. Obtain the solution of the following homogeneous equations:

$$4X_1 + 3X_2 - 2X_3 = 0$$
$$-5X_1 + 7X_2 + X_3 = 0$$
$$-12X_1 - 9X_2 + 6X_3 = 0$$

5.5-3. Obtain the solution of the following homogeneous equations:

$$4X_1 + X_2 + 2X_3 = 0$$
$$X_1 - 2X_2 - 4X_3 = 0$$
$$-3X_1 + 5X_2 + 10X_3 = 0$$

5.5-4. Obtain the solution of the following homogeneous equations:

$$2X_1 - 3X_2 + 8X_3 = 0$$
$$4X_1 + 3X_2 + 7X_3 = 0$$
$$2X_1 + 3X_2 + 2X_3 = 0$$

5.5-5. Obtain the solution of the following homogeneous equations:

$$2X_1 + 5X_2 - 2X_3 - 10X_4 = 0$$
$$-X_1 + 3X_2 + 2X_3 + 8X_4 = 0$$
$$5X_1 + 13X_2 + 3X_3 - X_4 = 0$$
$$-5X_1 - 2X_2 - X_3 + 7X_4 = 0$$

5.5-6. Obtain the solution of the following homogeneous equations:

$$2X_1 + 3X_2 + X_3 - 3X_4 = 0$$
$$-X_1 + X_2 - 4X_3 + 3X_4 = 0$$
$$3X_1 + 6X_2 - X_3 - 4X_4 = 0$$
$$6X_1 + 11X_2 - X_3 - 9X_4 = 0$$

5.6-1. Determine whether the following set of vectors is linearly dependent or independent:

$$\mathbf{A}_1 = [3 \quad 5] \qquad \mathbf{A}_2 = [-2 \quad 7] \qquad \mathbf{A}_3 = [8 \quad 3]$$

5.6-2. Determine whether the following set of vectors is linearly dependent or independent:

$$\mathbf{B}_1 = [\ 2 \quad -3] \qquad \mathbf{B}_2 = [2 \quad -7]$$
$$\mathbf{B}_3 = [-3 \quad 4] \qquad \mathbf{B}_4 = [1 \quad -10]$$

5.6–3. Determine whether the following set of vectors is linearly dependent or independent:

$$\mathbf{C}_1 = [-2 \quad 1 \quad 4] \qquad \mathbf{C}_2 = [3 \quad 5 \quad -2]$$
$$\mathbf{C}_3 = [-7 \quad -3 \quad 8]$$

5.6–4. Determine whether the following set of vectors is linearly dependent or independent:

$$\mathbf{D}_1 = [2 \quad 0 \quad 1 \quad -5] \qquad \mathbf{D}_2 = [6 \quad 3 \quad -1 \quad -4]$$
$$\mathbf{D}_3 = [0 \quad -3 \quad 4 \quad -11]$$

5.6–5. Find the value of x so that the three vectors are linearly dependent:

$$\mathbf{E}_1 = [5 \quad 2 \quad 5] \qquad \mathbf{E}_2 = [-3 \quad 1 \quad 0]$$
$$\mathbf{E}_3 = [-5 \quad 9 \quad x]$$

5.6–6. Express the vector $\mathbf{F}_1$ as a linear combination of $\mathbf{F}_2$, $\mathbf{F}_3$, and $\mathbf{F}_4$:

$$\mathbf{F}_1 = [2 \quad 0 \quad -1] \qquad \mathbf{F}_2 = [1 \quad 3 \quad 0]$$
$$\mathbf{F}_3 = [1 \quad 4 \quad 1] \qquad \mathbf{F}_4 = [1 \quad -2 \quad 0]$$

5.6–7. Express the vector $\mathbf{G}_1$ as a linear combination of $\mathbf{G}_2$ and $\mathbf{G}_3$:

$$\mathbf{G}_1 = [4 \quad -1 \quad 7 \quad 2] \qquad \mathbf{G}_2 = [-4 \quad 15 \quad -13 \quad 14]$$
$$\mathbf{G}_3 = [2 \quad 3 \quad 2 \quad 5]$$

5.7–1. What is the rank of a row or column vector that is not null?

5.7–2. What is the rank of an identity matrix of order n?

5.7–3. What is the rank of a square matrix of order n having all elements the same (but not zero)? What is the rank of a rectangular matrix of order $m \times n$ if all elements are the same (but not zero)?

5.7–4. Find the rank of each of the following matrices:

$$\mathbf{A} = \begin{bmatrix} 2 & 4 & -1 \\ 1 & 2 & 3 \\ 4 & 3 & 5 \end{bmatrix} \qquad \mathbf{B} = \begin{bmatrix} 5 & 2 & 7 \\ 1 & -4 & 3 \\ 11 & 0 & 17 \end{bmatrix} \qquad \mathbf{C} = \begin{bmatrix} 12 & 24 & -15 \\ 8 & 16 & -10 \\ -4 & -8 & 5 \end{bmatrix}$$

5.7–5. Find the rank of each of the following matrices:

$$\mathbf{D} = \begin{bmatrix} 3 & -2 & 4 & 1 \\ 5 & 6 & 2 & -3 \\ -2 & -8 & 2 & 5 \end{bmatrix} \qquad \mathbf{E} = \begin{bmatrix} 4 & -1 & 3 & 0 \\ -14 & 17 & -6 & 9 \\ -2 & 5 & 0 & 3 \end{bmatrix}$$

5.8–1. Determine the nature of the solution (see Table 5–1) of the following simultaneous equations, solve the equations, and express the result in the form of a vector $\mathbf{X}$:

$$X_1 + 3X_2 = -17$$
$$6X_1 - 2X_2 = 18$$

5.8-2. Solve the preceding problem for the following equations:

$$-3X_1 + 4X_2 = -2$$
$$9X_1 - 12X_2 = 8$$

5.8-3. Solve Problem 5.8-1 for the following equations:

$$5X_1 - 2X_2 = 3$$
$$-15X_1 + 6X_2 = -9$$

5.8-4. Solve Problem 5.8-1 for the following equations:

$$-3X_1 + 4X_2 = 0$$
$$2X_1 + 3X_2 = 0$$

5.8-5. Solve Problem 5.8-1 for the following equations:

$$6X_1 + X_2 = 0$$
$$-12X_1 - 2X_2 = 0$$

5.8-6. Two nonhomogeneous simultaneous equations have the general form

$$aX_1 + bX_2 = c$$
$$dX_1 + eX_2 = f$$

(a) Under what conditions do these equations have no solution?

(b) Under what conditions do these equations have only one solution?

(c) Under what conditions do these equations have an infinite number of solutions?

5.8-7. Solve Problem 5.8-1 for the following equations:

$$X_1 - X_2 + X_3 = 4$$
$$-2X_1 + X_2 + 2X_3 = 1$$
$$3X_1 - 2X_2 + X_3 = 7$$

5.8-8. Solve Problem 5.8-1 for the following equations:

$$3X_1 - 2X_2 + 4X_3 = -1$$
$$6X_1 + X_2 - 5X_3 = 11$$
$$X_1 + X_2 - 3X_3 = 4$$

5.8-9. Solve Problem 5.8-1 for the following equations:

$$-X_1 + 2X_2 + 4X_3 = -3$$
$$4X_1 - 8X_2 - 16X_3 = 12$$
$$-2X_1 + 4X_2 + 8X_3 = -6$$

5.8-10. Solve Problem 5.8-1 for the following equations:

$$3X_1 - 2X_2 + X_3 = -2$$
$$-X_1 + 3X_2 - 2X_3 = 5$$
$$2X_1 + X_2 - X_3 = 4$$

5.8-11. Solve Problem 5.8-1 for the following equations:

$$X_1 + 2X_2 - 3X_3 = 4$$
$$-2X_1 - 4X_2 + 6X_3 = 2$$
$$3X_1 + 6X_2 - 9X_3 = -1$$

5.8-12. Solve Problem 5.8-1 for the following equations:

$$3X_1 + 5X_2 - 2X_3 = 0$$
$$X_1 - 7X_2 + X_3 = 0$$
$$-2X_1 + X_2 - 4X_3 = 0$$

5.8-13. Solve Problem 5.8-1 for the following equations:

$$5X_1 - 2X_2 + 3X_3 = 0$$
$$2X_1 + 7X_2 - 3X_3 = 0$$
$$-X_1 + 3X_2 - 2X_3 = 0$$

5.8-14. Solve Problem 5.8-1 for the following equations:

$$2X_1 + 3X_2 - X_3 = 0$$
$$-6X_1 - 9X_2 + 3X_3 = 0$$
$$4X_1 + 6X_2 - 2X_3 = 0$$

6

...Eigenvalue Problems...

Eigenvalue problems arise in engineering analyses in a variety of situations, including the study of vibrations of elastic systems, buckling of structures, and oscillations of electrical networks. In this chapter several topics of importance in eigenvalue problems are presented.

6.1. Eigenvalues of a Matrix. Many engineering problems lead to a set of simultaneous equations that can be expressed in the matrix form

$$\mathbf{AX} = \lambda\mathbf{X} \tag{6-1}$$

in which **A** is a square matrix, **X** is a vector of unknowns, and λ is a scalar quantity. Equation (6-1) represents the *eigenvalue problem*, in which the solution of the simultaneous equations is a vector **X** such that when **X** is premultiplied by **A**, the product is a scalar multiple of **X** itself. Equation (6-1) can be rewritten as a homogeneous equation:

$$(\mathbf{A} - \lambda\mathbf{I})\mathbf{X} = \mathbf{0} \tag{6-2}$$

in which **I** is the identity matrix of the same order as the matrix **A**. When written in detailed form, the preceding equations appear as follows:

$$\begin{aligned}
(A_{11} - \lambda)X_1 + A_{12}X_2 + \ldots + A_{1n}X_n &= 0 \\
A_{21}X_1 + (A_{22} - \lambda)X_2 + \ldots + A_{2n}X_n &= 0 \\
\ldots \qquad \ldots \qquad \ldots \qquad \ldots & \\
A_{n1}X_1 + A_{n2}X_2 + \ldots + (A_{nn} - \lambda)X_n &= 0
\end{aligned} \tag{6-3}$$

Because Equations (6-3) are homogeneous, they are consistent and always have a trivial solution (**X** = **0**). However, it is the nontrivial solutions that are of interest, and they exist only if the determinant of the coefficient matrix is zero. Therefore, the following equation is obtained:

$$|\mathbf{A}-\lambda\mathbf{I}| = \begin{vmatrix} A_{11}-\lambda & A_{12} & \dots & A_{1n} \\ A_{21} & A_{22}-\lambda & \dots & A_{2n} \\ \dots & \dots & \dots & \dots \\ A_{n1} & A_{n2} & \dots & A_{nn}-\lambda \end{vmatrix} = 0 \qquad (6\text{–}4)$$

Equation (6–4) expresses the condition that is required for the existence of a nontrivial solution of the simultaneous equations and is called the *characteristic equation*. Usually the coefficients A_{ij} of the matrix **A** are known, and Equation (6–4) provides a means of determining the values of λ in order that a nontrivial solution will exist. Such values of λ are called *eigenvalues*,* and the corresponding solutions for **X** are called *eigenvectors*.†

If the determinant in Equation (6–4) is expanded, a polynomial expression in terms of λ is obtained. The highest power term in this *characteristic polynomial* is λ^n. Equating the polynomial to zero gives the following general form of the characteristic equation:

$$b_n\lambda^n + b_{n-1}\lambda^{n-1} + \dots + b_1\lambda + b_0 = 0 \qquad (6\text{–}5)$$

The coefficients in this equation have known values, since they depend only upon the elements A_{ij} of the matrix **A**. In particular, the coefficient b_n of λ^n is equal to $(-1)^n$, inasmuch as the first term in Equation (6–5) arises from the product of the principal diagonal elements in Equation (6–4). The other coefficients in Equation (6–5) can be expressed in terms of the principal minor determinants of **A**, but in general this representation is not very useful. However, it should be noted that the term b_0 is equal to the determinant of **A** and the coefficient b_{n-1} is equal to $(-1)^{n-1}$ times the sum of the elements on the principal diagonal of **A** (this sum is called the trace of the matrix; see Section 6.4).

The eigenvalues of the matrix **A** are the n roots of the characteristic equation (6–5). These roots may be real or complex, and some of them may be repeated. An eigenvalue that is different from all of the others is called a *distinct* eigenvalue; an eigenvalue that is repeated k times is said to be of *multiplicity* k. Matrices with repeated eigenvalues are discussed in Section 6.3.

Only matrices having real eigenvalues are discussed in this book. All symmetric matrices have real eigenvalues, but the eigenvalues of an unsymmetric matrix may be real or complex. Fortunately, in many engineering analyses involving unsymmetric matrices the eigenvalues turn out to be real because they represent physical quantities. In general, the real eigenvalues of a matrix may be positive, negative, or zero.

---

*Eigenvalues are referred to by many other names in mathematical literature; these names include *characteristic values, latent roots, proper values,* and *principal values.*

†Other names for eigenvectors are *characteristic vectors, proper vectors,* and *principal vectors.*

As an illustration of the calculation of the eigenvalues for a matrix, consider the following 2 X 2 matrix:

$$\mathbf{A} = \begin{bmatrix} 17 & -6 \\ 45 & -16 \end{bmatrix} \tag{6-6}$$

The first step in finding the eigenvalues is to obtain the matrix **(A − λI)** which appears in Equation (6–2):

$$(\mathbf{A} - \lambda\mathbf{I}) = \begin{bmatrix} 17-\lambda & -6 \\ 45 & -16-\lambda \end{bmatrix}$$

Next, the determinant of this matrix is equated to zero, as shown in Equation (6–4), to give the following relation:

$$\begin{vmatrix} 17-\lambda & -6 \\ 45 & -16-\lambda \end{vmatrix} = 0$$

When the determinant is expanded, the characteristic equation takes the form

$$\lambda^2 - \lambda - 2 = 0$$

or

$$(\lambda - 2)(\lambda + 1) = 0$$

Thus, the eigenvalues of the matrix **A**, which are the two roots of the characteristic equation, are

$$\lambda_1 = 2 \qquad \lambda_2 = -1 \tag{6-7}$$

The order in which these eigenvalues are selected is arbitrary, but for consistency the following rule will usually be adopted. The eigenvalues are taken in sequence according to their algebraic values with the one having the largest algebraic value being first. (An exception to this rule is in Section 6.9, where the eigenvalues are taken in sequence according to their absolute values.)

As a second example, the eigenvalues of the following third-order matrix will be determined:

$$\mathbf{B} = \begin{bmatrix} 20 & -4 & 8 \\ -40 & 8 & -20 \\ -60 & 12 & -26 \end{bmatrix} \tag{6-8}$$

The characteristic equation is obtained from Equation (6–4):

$$\begin{vmatrix} 20-\lambda & -4 & 8 \\ -40 & 8-\lambda & -20 \\ -60 & 12 & -26-\lambda \end{vmatrix} = 0$$

Expansion of this third-order determinant produces the polynomial equation:

$$-\lambda^3 + 2\lambda^2 + 8\lambda = 0$$

The three roots of this equation can be found by any one of several methods for solving a cubic equation, including trial and error. In this case, the polynomial can be factored (after multiplying through by minus one), as follows:

$$(\lambda - 4)(\lambda - 0)(\lambda + 2) = 0$$

which allows a direct determination of the roots. Thus, the eigenvalues of the matrix **B** are

$$\lambda_1 = 4 \qquad \lambda_2 = 0 \qquad \lambda_3 = -2 \tag{6-9}$$

Typically, the characteristic equation cannot be factored as easily as in this example, and the task of ascertaining the characteristic roots is much more arduous. For this reason, an iterative technique for finding the eigenvalues and eigenvectors of a matrix is given in Section 6.9.

6.2. Eigenvectors of a Matrix. After the eigenvalues of a matrix have been determined, the associated eigenvectors can be obtained by solving the appropriate sets of homogeneous equations (see Equations 6-2 or 6-3). Corresponding to each distinct eigenvalue λ_i, there will be a nonzero eigenvector $\mathbf{X}_i$. This eigenvector is the solution of the homogeneous equations that are obtained by substituting the value of λ_i into Equation (6-2), as follows:

$$(\mathbf{A} - \lambda_i \mathbf{I})\mathbf{X}_i = \mathbf{0} \tag{6-10}$$

To show how the eigenvectors can be determined in a numerical example, consider again the 2 X 2 matrix **A** given in the preceding section (see Equation 6-6). The first eigenvalue is $\lambda_1 = 2$ (see Equations 6-7), and when this value is substituted, Equation (6-10) becomes

$$(\mathbf{A} - \lambda_1 \mathbf{I})\mathbf{X}_1 = \begin{bmatrix} 17-2 & -6 \\ 45 & -16-2 \end{bmatrix} \begin{bmatrix} X_1 \\ X_2 \end{bmatrix} = \begin{bmatrix} 0 \\ 0 \end{bmatrix}$$

Therefore, the homogeneous equations to be solved (see Equations 6-3) are:

$$15X_1 - 6X_2 = 0$$

$$45X_1 - 18X_2 = 0$$

These equations are linearly dependent, and the coefficient matrix $(\mathbf{A} - \lambda_1 \mathbf{I})$ has rank equal to one. Therefore, the solution of the equations consists of one linearly independent vector (see Table 5-1) which can be found as described in Sections 5.5 and 5.8:

$$\mathbf{X}_1 = \beta_1 \begin{bmatrix} 2 \\ 5 \end{bmatrix} \tag{a}$$

in which β_1 is a nonzero arbitrary constant.

The second eigenvector is determined by substituting the second eigenvalue ($\lambda_2 = -1$) into Equation (6–10) and solving. The result is

$$\mathbf{X}_2 = \beta_2 \begin{bmatrix} 1 \\ 3 \end{bmatrix} \tag{b}$$

in which β_2 is arbitrary (but not zero).

The preceding solutions for the eigenvalues and eigenvectors can be verified by substitution into Equation (6–1). For the first eigenvalue and associated eigenvector, the equation is

$$\mathbf{AX}_1 = \begin{bmatrix} 17 & -6 \\ 45 & -16 \end{bmatrix} \beta_1 \begin{bmatrix} 2 \\ 5 \end{bmatrix} = \beta_1 \begin{bmatrix} 4 \\ 10 \end{bmatrix} = 2\beta_1 \begin{bmatrix} 2 \\ 5 \end{bmatrix} = \lambda_1 \mathbf{X}_1$$

which provides a check on the results. The same kind of check can be repeated for λ_2 and $\mathbf{X}_2$.

The n eigenvectors of a matrix can be arranged as the columns of a square matrix, which is called the *modal matrix*. Thus, if the eigenvectors of a matrix $\mathbf{A}$ are denoted as $\mathbf{X}_1, \mathbf{X}_2, \ldots, \mathbf{X}_n$, the modal matrix $\mathbf{M}$ is

$$\mathbf{M} = [\mathbf{X}_1 \quad \mathbf{X}_2 \quad \ldots \quad \mathbf{X}_n] \tag{6–11}$$

and has the same order as the original square matrix $\mathbf{A}$. There is an infinite number of modal matrices for each matrix $\mathbf{A}$, because each eigenvector $\mathbf{X}_i$ has a scalar multiplier β_i that can be assigned an arbitrary nonzero value.

For uniformity in constructing the modal matrix $\mathbf{M}$, the eigenvectors may be scaled according to some established rule. For instance, the first element may be made equal to unity. (If the first element is zero, the next element is made equal to unity, and so forth.) Another possibility is to make the numerically largest element in each eigenvector equal to unity. Both of these schemes determine the eigenvectors in a unique manner, and hence the modal matrix $\mathbf{M}$ also is unique.

A common way of scaling the eigenvectors is to normalize them by making their magnitudes equal to unity (see Section 1.5). In this case, each eigenvector is a unit vector and is determined uniquely except for a scalar factor of ± 1. To remove this ambiguity, an additional rule is needed; for example, the first nonzero element in the vector can be made positive. When the eigenvectors are scaled in this manner, they are called *normalized eigenvectors*. Such vectors will be denoted by lower case letters $\mathbf{x}_1, \mathbf{x}_2, \ldots, \mathbf{x}_n$, and the modal matrix of normalized eigenvectors will be denoted by the symbol $\mathbf{N}$. Thus,

$$\mathbf{N} = [\mathbf{x}_1 \quad \mathbf{x}_2 \quad \ldots \quad \mathbf{x}_n] \tag{6–12}$$

Note that the sequence in which the eigenvectors appear in the model matrices $\mathbf{M}$ and $\mathbf{N}$ is the same as the sequence of the associated eigenvalues.

If the constants β_1 and β_2 in the preceding example are selected as unity (see Equations a and b), the modal matrix of the matrix **A** is

$$\mathbf{M} = \begin{bmatrix} 2 & 1 \\ 5 & 3 \end{bmatrix}$$

If the eigenvectors are scaled so that the first elements are unity, the modal matrix is

$$\mathbf{M} = \begin{bmatrix} 1 & 1 \\ 2.5 & 3 \end{bmatrix}$$

Finally, if the eigenvectors are normalized to unit length, the following modal matrix of normalized eigenvectors is obtained:

$$\mathbf{N} = \begin{bmatrix} \dfrac{2}{\sqrt{29}} & \dfrac{1}{\sqrt{10}} \\ \dfrac{5}{\sqrt{29}} & \dfrac{3}{\sqrt{10}} \end{bmatrix}$$

Note that each column of this matrix is an eigenvector of unit magnitude.

Another matrix of great importance is the *spectral matrix*; this matrix (denoted by the symbol **S**) is a diagonal matrix having the eigenvalues along the principal diagonal:

$$\mathbf{S} = \begin{bmatrix} \lambda_1 & 0 & \dots & 0 \\ 0 & \lambda_2 & \dots & 0 \\ \dots & \dots & \dots & \dots \\ 0 & 0 & \dots & \lambda_n \end{bmatrix} \tag{6-13}$$

As an illustration, the spectral matrix for the preceding numerical example is

$$\mathbf{S} = \begin{bmatrix} 2 & 0 \\ 0 & -1 \end{bmatrix}$$

Both the modal matrix and the spectral matrix will be needed in later discussions.

The second example of the preceding article dealt with a third-order matrix **B** (see Equation 6-8) having eigenvalues given by Equations (6-9). The homogeneous equations corresponding to λ_1 in that example are

$$16X_1 - 4X_2 + 8X_3 = 0$$

$$-40X_1 + 4X_2 - 20X_3 = 0$$

$$-60X_1 + 12X_2 - 30X_3 = 0$$

These equations have rank equal to two, and hence their solution consists of one linearly independent solution vector with one arbitrary constant (see Table 5-1). This solution is the first eigenvector:

$$\mathbf{X}_1 = \beta_1 \begin{bmatrix} 1 \\ 0 \\ -2 \end{bmatrix} \tag{c}$$

The remaining two eigenvectors can be found in the same way:

$$\mathbf{X}_2 = \beta_2 \begin{bmatrix} 1 \\ 5 \\ 0 \end{bmatrix} \qquad \mathbf{X}_3 = \beta_3 \begin{bmatrix} 0 \\ 2 \\ 1 \end{bmatrix} \tag{d}$$

Thus, the modal matrix (for $\beta_1 = \beta_2 = \beta_3 = 1$) and the spectral matrix are

$$\mathbf{M} = \begin{bmatrix} 1 & 1 & 0 \\ 0 & 5 & 2 \\ -2 & 0 & 1 \end{bmatrix} \qquad \mathbf{S} = \begin{bmatrix} 4 & 0 & 0 \\ 0 & 0 & 0 \\ 0 & 0 & -2 \end{bmatrix}$$

Also, the matrix of normalized eigenvectors is

$$\mathbf{N} = \begin{bmatrix} \frac{1}{\sqrt{5}} & \frac{1}{\sqrt{26}} & 0 \\ 0 & \frac{5}{\sqrt{26}} & \frac{2}{\sqrt{5}} \\ -\frac{2}{\sqrt{5}} & 0 & \frac{1}{\sqrt{5}} \end{bmatrix}$$

This matrix is determined uniquely in accordance with the convention previously explained.

The eigenvectors of a matrix are linearly independent when they are associated with distinct eigenvalues, as in the preceding example. The independence of the eigenvectors $\mathbf{X}_1$, $\mathbf{X}_2$, and $\mathbf{X}_3$ (see Equations c and d) can be verified easily by testing them with Equation (5-13) of Section 5.6. It will be found that the only solution for the α's in that equation is the trivial one.

If a matrix of order n has n distinct eigenvalues, it follows that it will have n linearly independent eigenvectors. Therefore, the modal matrix $\mathbf{M}$ will have rank n, which means that it is nonsingular and has an inverse. Of course, if some of the eigenvectors associated with repeated eigenvalues are linearly dependent, then the corresponding modal matrix will be singular with rank less than n. The eigenvectors of a symmetric matrix have special properties that are described later in Section 6.7.

It was stated in the preceding section that only matrices having real eigenvalues are considered in this book. Because the existence of real eigenvalues means that all of the coefficients of the homogeneous equations (see Equation 6-3) are real, it follows that these equations have real solutions. Therefore, the eigenvectors of a matrix having real eigenvalues can be chosen to be real, as in the preceding numerical examples. (Because the constant β associated with each eigenvector is arbitrary, it is also possible to choose imaginary or complex eigenvectors.)

6.3. Repeated Eigenvalues. The eigenvalues of a matrix may be distinct or repeated, as explained in Section 6.1. Each distinct eigenvalue has a corresponding eigenvector that is linearly independent of all of the other eigenvectors. A repeated eigenvalue has as many eigenvectors as the multiplicity of the eigenvalue; for instance, an eigenvalue that is repeated three times has three eigenvectors associated with it. The eigenvectors associated with a repeated eigenvalue may be linearly dependent or independent; of course, each one is independent of the eigenvectors corresponding to any distinct eigenvalues.

In general, if an eigenvalue λ_i of an unsymmetric matrix has multiplicity k, the number of corresponding eigenvectors is also k. All or some of these eigenvectors may form a linearly independent set of vectors. For instance, all k eigenvectors may be linearly independent; or $k - 1$ vectors may be independent and the remaining vector dependent; or $k - 2$ vectors may be independent and two vectors dependent; and so forth, to the point where only one eigenvector is independent and the remaining $k - 1$ eigenvectors are dependent upon it.

The actual number of linearly independent eigenvectors associated with a repeated eigenvalue λ_i can be determined from the rank r of the coefficient matrix $(\mathbf{A} - \lambda_i\mathbf{I})$ of the homogeneous equations corresponding to the eigenvalue λ_i (see Equation 6-10). This number is equal to $n - r$ (see Table 5-1), in which n is the number of equations. Of course, the number $n - r$ must have a value from one to k; that is, $n - r = 1, 2, \ldots, k$. If all k eigenvectors are linearly independent, and if the other $n - k$ eigenvectors are associated with distinct eigenvalues (and hence also are linearly independent), then the original matrix will have n linearly independent eigenvectors. If the number of independent eigenvectors associated with the repeated eigenvalue is less than k, the original matrix will have fewer than n linearly independent eigenvectors.

The eigenvectors associated with a repeated eigenvalue are not necessarily determined uniquely, as they are for distinct eigenvalues. (In this context, uniqueness refers to the relative values of the elements in the eigenvectors, not the arbitrary constant β.) For instance, suppose that an eigenvalue has multiplicity two and that there are two linearly independent eigenvectors associated with that eigenvalue. Then there exists an infinite number of different sets of two eigenvectors, because new sets of two linearly independent eigenvectors can always be formed as linear combinations of previously selected sets.

All of the preceding ideas pertaining to repeated eigenvalues are illustrated in the following examples.

Example 1. Determine the eigenvalues and eigenvectors of the following matrix:

$$\mathbf{A} = \begin{bmatrix} 78 & -60 & 15 \\ 150 & -117 & 30 \\ 200 & -160 & 43 \end{bmatrix} \tag{6-14}$$

The first step in the solution is to write the homogeneous equations (see Equations 6-3):

$$(78 - \lambda)X_1 - 60X_2 + 15X_3 = 0 \tag{6-15a}$$

$$150X_1 + (-117 - \lambda)X_2 + 30X_3 = 0 \tag{6-15b}$$

$$200X_1 - 160X_2 + (43 - \lambda)X_3 = 0 \tag{6-15c}$$

From the determinant of these equations the characteristic equation is obtained:

$$-\lambda^3 + 4\lambda^2 + 3\lambda - 18 = 0$$

or

$$(\lambda - 3)(-\lambda + 3)(\lambda + 2) = 0$$

Hence, the eigenvalues of **A** are

$$\lambda_1 = 3 \qquad \lambda_2 = 3 \qquad \lambda_3 = -2 \tag{6-16}$$

and the spectral matrix is

$$\mathbf{S} = \begin{bmatrix} 3 & 0 & 0 \\ 0 & 3 & 0 \\ 0 & 0 & -2 \end{bmatrix} \tag{6-17}$$

Note that the first eigenvalue has multiplicity two.

The eigenvectors corresponding to the repeated eigenvalue are found by substituting $\lambda = 3$ into Equations (6-15) and solving:

$$75X_1 - 60X_2 + 15X_3 = 0$$

$$150X_1 - 120X_2 + 30X_3 = 0$$

$$200X_1 - 160X_2 + 40X_3 = 0$$

These equations have rank $r = 1$, so their solution consists of two linearly independent vectors (see Table 5-1). The solution is obtained by noting that the three equations are equivalent to the single equation

$$5X_1 - 4X_2 + X_3 = 0$$

Thus, the general form of the eigenvectors associated with the repeated eigenvalue is

$$\begin{bmatrix} 4X_2 - X_3 \\ 5X_2 \\ 5X_3 \end{bmatrix}$$

in which X_2 and X_3 can be selected arbitrarily. For instance, the following linearly independent eigenvectors may be chosen:

$$\mathbf{X}_1 = \beta_1 \begin{bmatrix} 4 \\ 5 \\ 0 \end{bmatrix} \qquad \mathbf{X}_2 = \beta_2 \begin{bmatrix} -1 \\ 0 \\ 5 \end{bmatrix} \tag{a}$$

However, as mentioned previously, these eigenvectors are not unique, and an infinite number of other choices is possible, such as

$$\mathbf{X}_1 = \beta_1 \begin{bmatrix} 3 \\ 5 \\ 5 \end{bmatrix} \qquad \mathbf{X}_2 = \beta_2 \begin{bmatrix} 7 \\ 10 \\ 5 \end{bmatrix} \tag{b}$$

Of course, each of these alternate eigenvectors is obtained as a linear combination of the first two (Equations a). It is even possible to select orthogonal eigenvectors, such as:

$$\mathbf{X}_1 = \beta_1 \begin{bmatrix} 1 \\ 1 \\ -1 \end{bmatrix} \qquad \mathbf{X}_2 = \beta_2 \begin{bmatrix} 1 \\ 2 \\ 3 \end{bmatrix}$$

Note that the scalar product of these vectors is zero.

The third eigenvalue ($\lambda_3 = -2$) yields the following homogeneous equations (see Equations 6-15):

$$80X_1 - 60X_2 + 15X_3 = 0$$

$$150X_1 - 115X_2 + 30X_3 = 0$$

$$200X_1 - 160X_2 + 45X_3 = 0$$

The rank r of these equations is two, so their solution contains one linearly independent vector (see Table 5-1). Solving the equations by elimination yields the following general form for the third eigenvector:

$$\begin{bmatrix} 3X_3 \\ 6X_3 \\ 8X_3 \end{bmatrix}$$

in which X_3 is arbitrary. Thus, the third eigenvector is

$$\mathbf{X}_3 = \beta_3 \begin{bmatrix} 3 \\ 6 \\ 8 \end{bmatrix} \tag{c}$$

and this eigenvector is unique (except for the scaling factor). Altogether, three linearly independent eigenvectors have been obtained for the matrix **A**.

Modal matrices **M** can now be constructed, such as the following (see Equations a, b, and c):

$$\mathbf{M}_1 = \begin{bmatrix} 4 & -1 & 3 \\ 5 & 0 & 6 \\ 0 & 5 & 8 \end{bmatrix} \qquad \mathbf{M}_2 = \begin{bmatrix} 3 & 7 & 3 \\ 5 & 10 & 6 \\ 5 & 5 & 8 \end{bmatrix} \tag{6-18}$$

Of course, any number of other modal matrices can be formed by using different eigenvectors in the first two columns. The rank of these matrices is three, because the columns are linearly independent; hence, the modal matrices can be inverted.

Example 2. Determine the eigenvalues and eigenvectors of the following matrix:

$$\mathbf{B} = \begin{bmatrix} 1 & 6 & 3 \\ 0 & 3 & 1 \\ -1 & 3 & 3 \end{bmatrix} \tag{6-19}$$

For this matrix, the homogeneous equations become (see Equations 6-3):

$$\begin{aligned} (1-\lambda)X_1 + 6X_2 + 3X_3 &= 0 \\ (3-\lambda)X_2 + X_3 &= 0 \\ -X_1 + 3X_2 + (3-\lambda)X_3 &= 0 \end{aligned} \tag{6-20}$$

and the characteristic equation reduces to

$$(\lambda-3)(\lambda-3)(\lambda-1) = 0$$

Thus, the eigenvalues are

$$\lambda_1 = 3 \qquad \lambda_2 = 3 \qquad \lambda_3 = 1$$

and the spectral matrix is

$$\mathbf{S} = \begin{bmatrix} 3 & 0 & 0 \\ 0 & 3 & 0 \\ 0 & 0 & 1 \end{bmatrix} \tag{6-21}$$

For the repeated eigenvalue the homogeneous equations (Equations 6-20) become

$$-2X_1 + 6X_2 + 3X_3 = 0$$
$$X_3 = 0$$
$$-X_1 + 3X_2 = 0$$

The rank of these equations is two, hence $n - r = 1$ and there is only one linearly independent eigenvector corresponding to the repeated eigenvalue. By solving the equations this eigenvector is found to be:

$$X_1 = \beta_1 \begin{bmatrix} 3 \\ 1 \\ 0 \end{bmatrix}$$

The second eigenvector corresponding to the repeated eigenvalue is the same as $\mathbf{X}_1$, except that a different scaling factor may be used if desired.

The eigenvector corresponding to the third eigenvalue is readily obtained from Equations (6-20) by substituting $\lambda_3 = 1$. The result is

$$\mathbf{X}_3 = \beta_3 \begin{bmatrix} 1 \\ -1 \\ 2 \end{bmatrix}$$

Because the matrix **B** has only two linearly independent eigenvectors, any modal matrix, such as

$$\mathbf{M} = \begin{bmatrix} 3 & 3 & 1 \\ 1 & 1 & -1 \\ 0 & 0 & 2 \end{bmatrix} \tag{6-22}$$

will have rank two. Therefore, **M** is singular and cannot be inverted.

Example 3. Investigate the eigenvalues and eigenvectors of a diagonal matrix with equal diagonal elements:

$$\mathbf{C} = \begin{bmatrix} 2 & 0 & 0 \\ 0 & 2 & 0 \\ 0 & 0 & 2 \end{bmatrix} \tag{6-23}$$

In this case, the homogeneous equations (Equations 6-3) become

$$(2-\lambda)X_1 = 0 \qquad (2-\lambda)X_2 = 0 \qquad (2-\lambda)X_3 = 0$$

and hence there is one eigenvalue of multiplicity three:

$$\lambda_1 = \lambda_2 = \lambda_3 = 2$$

Therefore, the spectral matrix is the same as the matrix **C** itself:

$$\mathbf{S} = \begin{bmatrix} 2 & 0 & 0 \\ 0 & 2 & 0 \\ 0 & 0 & 2 \end{bmatrix} \tag{6-24}$$

Substituting the eigenvalues into the preceding equations yields

$$(0)X_1 = 0 \qquad (0)X_2 = 0 \qquad (0)X_3 = 0$$

This set of equations has rank $r = 0$, which means that $n - r = 3$ and the matrix has three linearly independent eigenvectors. The general form of the eigenvectors is

$$\begin{bmatrix} X_1 \\ X_2 \\ X_3 \end{bmatrix}$$

in which X_1, X_2, and X_3 are arbitrary (except that all three cannot be zero). Therefore, the choice of eigenvectors is unlimited. However, the simplest set is

$$X_1 = \beta_1 \begin{bmatrix} 1 \\ 0 \\ 0 \end{bmatrix} \qquad X_2 = \beta_2 \begin{bmatrix} 0 \\ 1 \\ 0 \end{bmatrix} \qquad X_3 = \beta_3 \begin{bmatrix} 0 \\ 0 \\ 1 \end{bmatrix}$$

which leads to a modal matrix in the form of the identity matrix:

$$\mathbf{M} = \begin{bmatrix} 1 & 0 & 0 \\ 0 & 1 & 0 \\ 0 & 0 & 1 \end{bmatrix} \tag{6-25}$$

However, any nonsingular matrix of third order would be a valid modal matrix. Note that the matrix **C** has an eigenvalue of multiplicity three and three linearly independent eigenvectors.

Example 4. Investigate the eigenvalues and eigenvectors of the following lower triangular matrix:

$$\mathbf{D} = \begin{bmatrix} 2 & 0 & 0 \\ 3 & 2 & 0 \\ 4 & 0 & 2 \end{bmatrix} \tag{6-26}$$

This matrix has the same repeated eigenvalue as does the matrix **C** of the preceding example:

$$\lambda_1 = \lambda_2 = \lambda_3 = 2$$

The resulting homogeneous equations are

$$(0)X_1 = 0 \qquad 3X_1 + (0)X_2 = 0 \qquad 4X_1 + (0)X_3 = 0$$

with rank $r = 1$. Therefore, two linearly independent eigenvectors exist; their general form is

$$\begin{bmatrix} 0 \\ X_2 \\ X_3 \end{bmatrix}$$

with X_2 and X_3 arbitrary (but not both zero). One choice for the two linearly independent eigenvectors is as follows:

$$\mathbf{X}_1 = \beta_1 \begin{bmatrix} 0 \\ 1 \\ 0 \end{bmatrix} \qquad \mathbf{X}_2 = \beta_2 \begin{bmatrix} 0 \\ 0 \\ 1 \end{bmatrix}$$

but an infinite number of other choices exists. The third eigenvector is not independent and must be a linear combination of the first two; again, many possibilities exist, such as

$$\mathbf{X}_3 = \beta_3 \begin{bmatrix} 0 \\ 1 \\ 1 \end{bmatrix}$$

The modal matrix corresponding to the preceding eigenvectors is

$$\mathbf{M} = \begin{bmatrix} 0 & 0 & 0 \\ 1 & 0 & 1 \\ 0 & 1 & 1 \end{bmatrix} \tag{6-27}$$

All modal matrices for the matrix **D** are singular matrices of rank two because they contain only two linearly independent columns. In this case, the matrix **D** has an eigenvalue of multiplicity three, but only two linearly independent eigenvectors.

Example 5. Investigate the eigenvalues and eigenvectors of the following lower triangular matrix:

$$\mathbf{E} = \begin{bmatrix} 2 & 0 & 0 \\ 3 & 2 & 0 \\ 4 & 5 & 2 \end{bmatrix} \tag{6-28}$$

This matrix also has the same repeated eigenvalues:

$$\lambda_1 = \lambda_2 = \lambda_3 = 2$$

and the corresponding homogeneous equations are

$$(0)X_1 = 0 \qquad 3X_1 + (0)X_2 = 0 \qquad 4X_1 + 5X_2 + (0)X_3 = 0$$

with rank $r = 2$. Therefore, the matrix **E** has one linearly independent eigenvector of the general form

$$\begin{bmatrix} 0 \\ 0 \\ X_3 \end{bmatrix}$$

in which X_3 has any nonzero value. Therefore, the first eigenvector is

$$\mathbf{X}_1 = \beta_1 \begin{bmatrix} 0 \\ 0 \\ 1 \end{bmatrix}$$

The other two eigenvectors are the same except that the scalar factor may be different. Some possible modal matrices are

$$\mathbf{M}_1 = \begin{bmatrix} 0 & 0 & 0 \\ 0 & 0 & 0 \\ 1 & 1 & 1 \end{bmatrix} \qquad \mathbf{M}_2 = \begin{bmatrix} 0 & 0 & 0 \\ 0 & 0 & 0 \\ 1 & 2 & 3 \end{bmatrix}$$

These modal matrices have only one linearly independent column, so they are singular with rank equal to one. Note that the original matrix **E** has an eigenvalue of multiplicity three but only one linearly independent eigenvector.

Examples 3, 4, and 5 illustrate that the number of linearly independent eigenvectors corresponding to an eigenvalue of multiplicity k may be anywhere from 1 to k.

6.4. Properties of Eigenvalues and Eigenvectors. There are many properties and relationships pertaining to the eigenvalues and eigenvectors of a square matrix, and a few of the more important ones are mentioned in this section.

Transpose of a Matrix. The first case to be considered is that of a matrix **A** and its transpose $\mathbf{A}^\mathsf{T}$. The eigenvalues of **A** are found by solving Equation (6–4), while the eigenvalues of $\mathbf{A}^\mathsf{T}$ are found from the following equation:

$$|\mathbf{A}^\mathsf{T} - \lambda\mathbf{I}| = 0$$

However, since the matrix **I** is not changed by transposition, this equation can be written as

$$|(\mathbf{A} - \lambda\mathbf{I})^\mathsf{T}| = 0 \qquad \text{(a)}$$

If it is recalled that the value of a determinant and its transpose are the same (see Equation 3-5), then it is apparent that Equation (6–4) and Equation (a) must produce the same characteristic equation. Hence, the conclusion is reached that the eigenvalues of a matrix and its transpose are the same. The eigenvectors for $\mathbf{A}$ and $\mathbf{A}^\mathsf{T}$ may be different; however, they are orthogonal to one another when they correspond to different eigenvalues (this property is explained at the end of the next section).

Inverse of a Matrix. Assume that both sides of Equation (6–1) are premultiplied by the inverse of the matrix $\mathbf{A}$, thereby yielding the equation

$$\mathbf{A}^{-1}\mathbf{A}\mathbf{X} = \lambda\mathbf{A}^{-1}\mathbf{X} \qquad \text{or} \qquad \mathbf{X} = \lambda\mathbf{A}^{-1}\mathbf{X}$$

Upon dividing both sides of this last equation by λ, and also interchanging the two sides of the equation, it becomes

$$\mathbf{A}^{-1}\mathbf{X} = \frac{1}{\lambda}\mathbf{X}$$

which has the same form as Equation (6–1), except that $\mathbf{A}$ is replaced by $\mathbf{A}^{-1}$ and λ is replaced by $1/\lambda$. Therefore, it may be concluded that the eigenvalues of $\mathbf{A}^{-1}$ are the reciprocals of the eigenvalues of $\mathbf{A}$ and that both $\mathbf{A}$ and $\mathbf{A}^{-1}$ have the same eigenvectors. Furthermore, the eigenvectors correspond to the same eigenvalues; that is, the eigenvector of $\mathbf{A}$ corresponding to the eigenvalue λ_i is the eigenvector of $\mathbf{A}^{-1}$ corresponding to the eigenvalue $1/\lambda_i$.

Matrix Multiplied by a Constant. If the matrix $\mathbf{A}$ in Equation (6–1) is multiplied by a scalar α, the equation becomes

$$\alpha\mathbf{A}\mathbf{X} = \alpha\lambda\mathbf{X}$$

This equation also has the same form as Equation (6–1) except that $\mathbf{A}$ is replaced by $\alpha\mathbf{A}$ and λ is replaced by $\alpha\lambda$. Hence, if a matrix is multiplied by a constant, the eigenvalues are multiplied by the same constant, but the eigenvectors are not changed.

Orthogonal Matrix. It will be recalled that an orthogonal matrix has the interesting property that its inverse and its transpose are the same (see Section 4.6). As a preliminary matter to determining the eigenvalues of an orthogonal matrix, note that the transpose of Equation (6–1) leads to the following relationships:

$$(\mathbf{A}\mathbf{X})^\mathsf{T} = \mathbf{X}^\mathsf{T}\mathbf{A}^\mathsf{T} = \lambda\mathbf{X}^\mathsf{T} \tag{b}$$

Hence, the product $(\mathbf{A}\mathbf{X})^\mathsf{T}(\mathbf{A}\mathbf{X})$ can be written as

$$(\mathbf{A}\mathbf{X})^\mathsf{T}(\mathbf{A}\mathbf{X}) = \mathbf{X}^\mathsf{T}\mathbf{A}^\mathsf{T}\mathbf{A}\mathbf{X} = \mathbf{X}^\mathsf{T}\mathbf{X} \tag{c}$$

because the product $\mathbf{A}^\mathsf{T}\mathbf{A}$ is equal to the identity matrix when $\mathbf{A}$ is orthogonal. Another way of writing the same product can be obtained by using the last expression in Equation (b) as well as Equation (6–1); thus,

$$(\mathbf{AX})^{\mathsf{T}}(\mathbf{AX}) = \lambda \mathbf{X}^{\mathsf{T}} \lambda \mathbf{X} = \lambda^2 \mathbf{X}^{\mathsf{T}} \mathbf{X} \tag{d}$$

Comparing Equations (c) and (d) shows that

$$\lambda = \pm 1$$

provided that $\mathbf{X}^{\mathsf{T}}\mathbf{X}$ is not itself zero, which can be assured whenever the eigenvector $\mathbf{X}$ is real. Of course, the eigenvector can be chosen as real whenever the corresponding eigenvalue λ is real. Thus, it can be concluded that all real eigenvalues of an orthogonal matrix must be equal to $+1$ or -1.

Triangular Matrix. Next, consider a matrix $\mathbf{A}$ that is upper triangular, so that Equation (6–4) becomes

$$\begin{vmatrix} A_{11} - \lambda & A_{12} & \dots & A_{1n} \\ 0 & A_{22} - \lambda & \dots & A_{2n} \\ \dots & \dots & \dots & \dots \\ 0 & 0 & \dots & A_{nn} - \lambda \end{vmatrix} = 0$$

The expansion of this determinant yields

$$(A_{11} - \lambda)(A_{22} - \lambda) \dots (A_{nn} - \lambda) = 0 \tag{e}$$

which shows that the n eigenvalues are the n diagonal elements of the matrix. The same conclusion holds for a lower triangular matrix.

If a triangular matrix of order n has n distinct eigenvalues (and hence it also has n linearly independent eigenvectors), and if the eigenvalues are numbered in the order in which they appear on the principal diagonal, then the modal matrix will also be triangular (and either upper or lower, according to whether the original matrix is upper or lower, respectively). If there are repeated eigenvalues, other forms of the modal matrix are possible.

Diagonal Matrix. A diagonal matrix may be considered as a special case of a triangular matrix in which all off-diagonal elements are zero. Therefore, Equation (e) is also the characteristic equation for a diagonal matrix. Hence, the eigenvalues of a diagonal matrix are the diagonal elements.

Product of Two Matrices. Now consider the matrix product $\mathbf{AB}$, in which $\mathbf{A}$ and $\mathbf{B}$ are square matrices of order n and $\mathbf{A}$ is nonsingular. The relationship between this product and one of its eigenvalues λ_i is given by the equation

$$\mathbf{ABY}_i = \lambda_i \mathbf{Y}_i \tag{f}$$

in which $\mathbf{Y}_i$ is the eigenvector of $\mathbf{AB}$ corresponding to λ_i. Inserting the product $\mathbf{AA}^{-1}$ into both sides of this equation gives

$$\mathbf{ABAA}^{-1}\mathbf{Y}_i = \lambda_i \mathbf{AA}^{-1}\mathbf{Y}_i$$

Premultiplying by $\mathbf{A}^{-1}$ yields

$$\mathbf{BAA}^{-1}\mathbf{Y}_i = \lambda_i \mathbf{A}^{-1}\mathbf{Y}_i$$

If the vector $\mathbf{A}^{-1}\mathbf{Y}_i$ is denoted $\mathbf{Z}_i$, this equation becomes

$$\mathbf{BAZ}_i = \lambda_i \mathbf{Z}_i \tag{g}$$

A comparison of Equations (f) and (g) shows that the matrices **AB** and **BA** have the same eigenvalues. Furthermore, the ith eigenvector of **BA** is equal to $\mathbf{A}^{-1}$ times the ith eigenvector of **AB**.

Zero Eigenvalues. When the eigenvalue λ_i of a matrix **A** is equal to zero, Equation (6-10) simplifies to

$$\mathbf{AX}_i = \mathbf{0}$$

From this equation, it is apparent that the matrix **A** must be singular in order to have a nonzero eigenvector $\mathbf{X}_i$. The converse of this statement is also true; that is, if the matrix **A** is singular there must be at least one zero eigenvalue. To be more specific, if the rank of **A** is r, then at least $n - r$ eigenvalues of **A** are zero. For example, the third-order matrix **B** given by Equation (6-8) in Section 6.1 has one zero eigenvalue; it also is singular and of rank two.

A matrix with repeated zero eigenvalues is handled in the same manner as one with repeated nonzero eigenvalues (see Section 6.3). Also, an unsymmetric matrix may have all eigenvalues equal to zero even when the matrix itself is not a null matrix, although the only symmetric matrices having all zero eigenvalues are null matrices.

Trace of a Matrix. The sum of the elements on the principal diagonal of any matrix **A** is called the *trace* of **A** and is written tr(**A**). By a more detailed consideration of the characteristic equation than can be given here, it is possible to show that the trace of a matrix is equal to the sum of its eigenvalues; thus

$$\mathrm{tr}(\mathbf{A}) = \sum_{i=1}^{n} A_{ii} = \sum_{i=1}^{n} \lambda_i \tag{6-29}$$

Furthermore, the coefficient b_{n-1} of the second term in the characteristic equation (6-5) is equal to $(-1)^{n-1}$ times the sum of the eigenvalues. In addition, it can be proven that the product of the eigenvalues of a matrix is equal to the determinant of the matrix, as shown:

$$|\mathbf{A}| = \lambda_1 \lambda_2 \ldots \lambda_n \tag{6-30}$$

As mentioned earlier, this product is also equal to the term b_0 in the characteristic polynomial (see Equation 6-5).

Many of the properties of eigenvalues and eigenvectors covered in this section can be verified by the reader for the numerical examples given previously.

Other illustrations of these properties will be found in the problems at the end of the chapter. Symmetric matrices have particular properties that are discussed in Section 6.7.

6.5. Diagonalization of a Matrix. In Section 6.2 the modal matrix **M** was defined as a matrix for which the columns are the n eigenvectors of a matrix **A** (see Equation 6-11). The eigenvectors $X_1, X_2, \ldots, X_n$ contained in the modal matrix must satisfy the equations:

$$\mathbf{AX}_1 = \lambda_1 \mathbf{X}_1 \qquad \mathbf{AX}_2 = \lambda_2 \mathbf{X}_2 \qquad \ldots \qquad \mathbf{AX}_n = \lambda_n \mathbf{X}_n$$

in which $\lambda_1, \lambda_2, \ldots, \lambda_n$ are the eigenvalues of **A** (see Equation 6-1). Now suppose that the modal matrix is premultiplied by the matrix **A**, thereby giving the following expressions:

$$\begin{aligned} \mathbf{AM} &= \mathbf{A}[\mathbf{X}_1 \quad \mathbf{X}_2 \quad \ldots \quad \mathbf{X}_n] \\ &= [\mathbf{AX}_1 \quad \mathbf{AX}_2 \quad \ldots \quad \mathbf{AX}_n] \\ &= [\lambda_1 \mathbf{X}_1 \quad \lambda_2 \mathbf{X}_2 \quad \ldots \quad \lambda_n \mathbf{X}_n] \end{aligned} \qquad \text{(a)}$$

The last expression is the modal matrix **M** but with the first column multiplied by λ_1, the second column multiplied by λ_2, and so forth. Hence, it can be rewritten as the matrix **M** postmultiplied by a diagonal matrix having as it diagonal elements the eigenvalues $\lambda_1, \lambda_2, \ldots, \lambda_n$. In other words, the last expression is equal to the product **MS**, where **S** is the spectral matrix. Thus, Equation (a) becomes

$$\mathbf{AM} = \mathbf{MS} \qquad \text{(6-31)}$$

or, upon premultiplying both sides of this equation by the inverse of **M**,

$$\mathbf{M}^{-1}\mathbf{AM} = \mathbf{S} \qquad \text{(6-32)}$$

All of the matrices appearing in this equation are of order $n \times n$. The modal matrix **M** has columns that are the eigenvectors of **A**, and the product $\mathbf{M}^{-1}\mathbf{AM}$ produces the spectral matrix, that is, a diagonal matrix having as diagonal elements the eigenvalues of **A**. Of course, the matrix **N** of normalized eigenvectors is a special case of a modal matrix and may be used in place of **M** in any expression (such as Equation 6-32).

Equation (6-32) is said to represent the *diagonalization* of the matrix **A**, or the *transformation* of **A** to a diagonal matrix. It is an important relationship between a matrix **A** and its modal and spectral matrices and is useful when dealing with principal coordinates (which are discussed in Chapter 7). Equation (6-32) should not be viewed, however, as a practical means of calculating the matrix **S**, since the eigenvalues are required even before the modal matrix can be obtained.

Any square matrix **A** of order n can be transformed in the manner indicated by Equation (6-32) provided that the matrix has n linearly independent eigenvectors. There are two conditions under which a matrix is guaranteed to have n linearly independent eigenvectors: first, if the matrix has n distinct (or different) eigenvalues, and second, if the matrix is symmetric. Under either of these conditions, the matrix always can be diagonalized. If the matrix is not symmetric, and if some of the eigenvalues are repeated, the matrix may or may not be diagonalizable, depending upon whether or not there are n linearly independent eigenvectors.

The following examples include both diagonalizable and non-diagonalizable matrices. Also, the diagonalization of symmetric matrices is discussed more fully in Section 6.7.

Example 1. The matrix **A** shown below is to be diagonalized by means of Equation (6–32):

$$\mathbf{A} = \begin{bmatrix} 16 & -8 \\ 24 & -12 \end{bmatrix}$$

The eigenvalues and eigenvectors of **A** are found by the methods described in Sections 6.1 and 6.2:

$$\lambda_1 = 4 \qquad \lambda_2 = 0$$

$$\mathbf{X}_1 = \beta_1 \begin{bmatrix} 2 \\ 3 \end{bmatrix} \qquad \mathbf{X}_2 = \beta_2 \begin{bmatrix} 1 \\ 2 \end{bmatrix}$$

Note that the matrix **A** has two distinct eigenvalues; hence, it has two linearly independent eigenvectors. Before obtaining a modal matrix, it is necessary to assign values to the arbitrary constants β_1 and β_2. If these constants are both taken as unity, the modal matrix and its inverse are

$$\mathbf{M} = \begin{bmatrix} 2 & 1 \\ 3 & 2 \end{bmatrix} \qquad \mathbf{M}^{-1} = \begin{bmatrix} 2 & -1 \\ -3 & 2 \end{bmatrix}$$

Then the transformation shown in Equation (6–32) becomes

$$\mathbf{M}^{-1}\mathbf{A}\mathbf{M} = \begin{bmatrix} 2 & -1 \\ -3 & 2 \end{bmatrix} \begin{bmatrix} 16 & -8 \\ 24 & -12 \end{bmatrix} \begin{bmatrix} 2 & 1 \\ 3 & 2 \end{bmatrix} = \begin{bmatrix} 4 & 0 \\ 0 & 0 \end{bmatrix} = \mathbf{S}$$

Note that the elements on the principal diagonal of **S** are the two eigenvalues of **A**. Furthermore, the order in which the eigenvalues appear on the diagonal is the same as the sequence of the associated eigenvectors in the columns of the modal matrix. Any other values could have been assigned to the arbitrary constants β_1 and β_2, and the transformation would still be valid.

Example 2. Diagonalize the matrix **B** shown below:

$$\mathbf{B} = \begin{bmatrix} 78 & -60 & 15 \\ 150 & -117 & 30 \\ 200 & -160 & 43 \end{bmatrix}$$

This matrix was analyzed in detail in Example 1 of Section 6.3 (see Equation 6–14), and it was found to have an eigenvalue of multiplicity two (see Equation 6–17 for the spectral matrix). Nevertheless, the matrix has three linearly independent eigenvectors, hence it can be diagonalized by Equation (6–32). Because the eigenvectors are not unique, any number of different modal matrices can be constructed, two examples of which are given by Equations (6–18). If the first of those matrices is utilized, the modal matrix and its inverse become

$$\mathbf{M} = \begin{bmatrix} 4 & -1 & 3 \\ 5 & 0 & 6 \\ 0 & 5 & 8 \end{bmatrix} \qquad \mathbf{M}^{-1} = \frac{1}{5}\begin{bmatrix} 30 & -23 & 6 \\ 40 & -32 & 9 \\ -25 & 20 & -5 \end{bmatrix}$$

Equation (6–32) then gives

$$\mathbf{M}^{-1}\mathbf{BM} = \begin{bmatrix} 3 & 0 & 0 \\ 0 & 3 & 0 \\ 0 & 0 & -2 \end{bmatrix} = \mathbf{S}$$

as expected. Alternatively, the second modal matrix in Equation (6–18), along with its inverse, could be used in Equation (6–32) to diagonalize the matrix.

Example 3. State whether the following third-order matrices can be diagonalized:

Matrix **B** of Example 2, Section 6.3 (see Equation 6–19)

Matrix **D** of Example 4, Section 6.3 (see Equation 6–26)

Matrix **E** of Example 5, Section 6.3 (see Equation 6–28)

As shown in the earlier examples, none of these matrices has three linearly independent eigenvectors. Therefore, their modal matrices are singular, and diagonalization by Equation (6–32) is not possible.

As an incidental matter, the relationship between the inverse of a matrix and the eigenvalues and eigenvectors of the original matrix can be observed at this point. If both sides of Equation (6–31) are postmultiplied by the inverse of the modal matrix, the result is

$$\mathbf{A} = \mathbf{MSM}^{-1}$$

Then the inverse of **A** must be the product of the inverses of the matrices on the right-hand side, but in reverse order (see Equation 4–18):

$$\mathbf{A}^{-1} = \mathbf{MS}^{-1}\mathbf{M}^{-1} \tag{6–33}$$

Thus, if the modal matrix, its inverse, and the inverse of the spectral matrix have all been determined, the inverse of the original matrix **A** can be found by matrix multiplications. However, this method is not recommended as a general procedure for finding the inverse of a matrix, because the determination of $\mathbf{M}, \mathbf{M}^{-1}$, and $\mathbf{S}^{-1}$ requires more effort than a direct calculation of $\mathbf{A}^{-1}$ by one of the methods described in Chapter 4.

The relationship between the eigenvalues and eigenvectors of a matrix $\mathbf{A}$ and its transpose $\mathbf{A}^\mathsf{T}$ can also be obtained from Equation (6-31). When the transposes of both sides of that equation are taken, the result is

$$\mathbf{M}^\mathsf{T}\mathbf{A}^\mathsf{T} = \mathbf{S}^\mathsf{T}\mathbf{M}^\mathsf{T} = \mathbf{S}\mathbf{M}^\mathsf{T}$$

Premultiplying and postmultiplying the expressions on both sides of this equation by the inverse of the transpose of $\mathbf{M}$ gives

$$(\mathbf{M}^\mathsf{T})^{-1}\mathbf{M}^\mathsf{T}\mathbf{A}^\mathsf{T}(\mathbf{M}^\mathsf{T})^{-1} = (\mathbf{M}^\mathsf{T})^{-1}\mathbf{S}\mathbf{M}^\mathsf{T}(\mathbf{M}^\mathsf{T})^{-1}$$

or

$$\mathbf{A}^\mathsf{T}(\mathbf{M}^\mathsf{T})^{-1} = (\mathbf{M}^\mathsf{T})^{-1}\mathbf{S} \tag{6-34}$$

This equation has the same form as Equation (6-31) and, hence, it confirms the fact that both $\mathbf{A}$ and $\mathbf{A}^\mathsf{T}$ have the same spectral matrix. Furthermore, Equation (6-34) shows that the modal matrix for $\mathbf{A}^\mathsf{T}$ is $(\mathbf{M}^\mathsf{T})^{-1}$, where $\mathbf{M}$ is the modal matrix for $\mathbf{A}$. Thus, the columns of the matrix $(\mathbf{M}^\mathsf{T})^{-1}$ must be the eigenvectors of $\mathbf{A}^\mathsf{T}$. Since $(\mathbf{M}^\mathsf{T})^{-1} = (\mathbf{M}^{-1})^\mathsf{T}$, it follows that the rows of $\mathbf{M}^{-1}$ are the eigenvectors of $\mathbf{A}^\mathsf{T}$. Finally, because $\mathbf{M}^{-1}\mathbf{M} = \mathbf{I}$, it is seen that the row vectors of $\mathbf{M}^{-1}$ times the column vectors of $\mathbf{M}$ always give the Kronecker delta. Therefore, the eigenvectors of $\mathbf{A}$ and $\mathbf{A}^\mathsf{T}$ corresponding to different eigenvalues are orthogonal vectors.

6.6. Types of Matrix Transformations. In the process of diagonalizing a matrix, a transformation $\mathbf{M}^{-1}\mathbf{A}\mathbf{M}$ was encountered (see Equation 6-32). This type of matrix transformation is only one of several kinds that arise in engineering work, and some of the other important transformations are mentioned in this section.

An *equivalence transformation* is defined by the equation

$$\mathbf{B} = \mathbf{T}_1\mathbf{A}\mathbf{T}_2 \tag{6-35}$$

in which $\mathbf{T}_1$ and $\mathbf{T}_2$ are any nonsingular square matrices. The matrices $\mathbf{A}$ and $\mathbf{B}$, which may be either rectangular or square, are said to be *equivalent matrices*, and the operation $\mathbf{T}_1\mathbf{A}\mathbf{T}_2$ is said to be an equivalence transformation on the matrix $\mathbf{A}$. The transformation matrices $\mathbf{T}_1$ and $\mathbf{T}_2$, which do not have to be of the same order, can always be obtained as the products of elementary matrices, as described in Section 2.6. An elementary transformation does not change the rank of a matrix because it does not change the linear dependence of the rows or columns of the matrix; hence, equivalent matrices have the same rank. Also, equivalent matrices are of the same order, as can be seen by applying the conformability requirements for matrix multiplication to the product $\mathbf{T}_1\mathbf{A}\mathbf{T}_2$.

A *congruence transformation* is defined by the equation

$$\mathbf{B} = \mathbf{T}^{\mathrm{T}}\mathbf{A}\mathbf{T} \tag{6-36}$$

in which **A**, **B**, and **T** are square matrices of the same order. The matrices **A** and **B** are said to be *congruent matrices*, and the transformation matrix **T** may be any arbitrary, square, nonsingular matrix. It is apparent that a congruence transformation is a special case of an equivalence transformation; therefore, congruent matrices have the same rank. Since it is always possible to consider the transposed matrix as the original matrix, a congruence transformation is defined alternatively by the equation $\mathbf{B} = \mathbf{T}\mathbf{A}\mathbf{T}^{\mathrm{T}}$.

The type of transformation encountered in diagonalizing a matrix (see Section 6.5) is a *similarity transformation*, defined by the expression

$$\mathbf{B} = \mathbf{T}^{-1}\mathbf{A}\mathbf{T} \tag{6-37}$$

As in the case of a congruence transformation, the matrices **A**, **B**, and **T** must be square and have the same order. The transformation matrix **T** must be nonsingular, but otherwise it is arbitrary. *Similar matrices* are also equivalent matrices, which means that **A** and **B** must have the same rank. (Note that $\mathbf{T}\mathbf{A}\mathbf{T}^{-1}$ is also a similarity transformation on the matrix **A**.)

An important property concerning the eigenvalues and eigenvectors of similar matrices can now be observed. The eigenvalues and eigenvectors of the matrix **A** satisfy the equation

$$\mathbf{A}\mathbf{X} = \lambda\mathbf{X} \tag{6-1}$$

repeated

Premultiplying both sides of this equation by $\mathbf{T}^{-1}$, and also inserting the identity matrix on the left-hand side, gives

$$\mathbf{T}^{-1}\mathbf{A}\mathbf{I}\mathbf{X} = \lambda\mathbf{T}^{-1}\mathbf{X}$$

Replacing **I** by the product $\mathbf{T}\mathbf{T}^{-1}$ yields

$$\mathbf{T}^{-1}\mathbf{A}\mathbf{T}\mathbf{T}^{-1}\mathbf{X} = \lambda\mathbf{T}^{-1}\mathbf{X}$$

If the product $\mathbf{T}^{-1}\mathbf{X}$ is denoted as the vector **Y** and the product $\mathbf{T}^{-1}\mathbf{A}\mathbf{T}$ is denoted as the matrix **B**, this equation can be rewritten as

$$\mathbf{B}\mathbf{Y} = \lambda\mathbf{Y} \tag{6-38}$$

This equation has the same form as Equation (6-1), hence it can be concluded that the eigenvectors of **B** are related to those of **A** by the equation

$$\mathbf{Y} = \mathbf{T}^{-1}\mathbf{X} \tag{6-39}$$

Furthermore, comparing Equations (6-38) and (6-1) shows that the eigenvalues of **A** and **B** are the same. Therefore, similar matrices have the same eigenvalues, which also means that they have the same characteristic equation, the same trace, and equal determinants.

In the preceding section it was shown that for any matrix **A** and its transpose $\mathbf{A}^\mathsf{T}$ the following equations are valid:

$$\mathbf{AM} = \mathbf{MS} \qquad \text{(6-31) repeated}$$

$$\mathbf{A}^\mathsf{T}(\mathbf{M}^\mathsf{T})^{-1} = (\mathbf{M}^\mathsf{T})^{-1}\mathbf{S} \qquad \text{(6-34) repeated}$$

If both sides of the first of these equations are premultiplied by $\mathbf{M}^{-1}$ and both sides of the second equation are premultiplied by $\mathbf{M}^\mathsf{T}$, they become

$$\mathbf{M}^{-1}\mathbf{AM} = \mathbf{S} \quad \text{and} \quad \mathbf{M}^\mathsf{T}\mathbf{A}^\mathsf{T}(\mathbf{M}^\mathsf{T})^{-1} = \mathbf{S}$$

from which

$$\mathbf{M}^{-1}\mathbf{AM} = \mathbf{M}^\mathsf{T}\mathbf{A}^\mathsf{T}(\mathbf{M}^\mathsf{T})^{-1}$$

If both sides are now premultiplied by **M** and postmultiplied by $\mathbf{M}^{-1}$, the equation takes the form

$$\mathbf{A} = \mathbf{MM}^\mathsf{T}\mathbf{A}^\mathsf{T}(\mathbf{M}^\mathsf{T})^{-1}\mathbf{M}^{-1} \qquad \text{(a)}$$

Introducing the notation

$$\mathbf{T} = \mathbf{MM}^\mathsf{T} \qquad \mathbf{T}^{-1} = (\mathbf{M}^\mathsf{T})^{-1}\mathbf{M}^{-1}$$

means that Equation (a) becomes

$$\mathbf{A} = \mathbf{TA}^\mathsf{T}\mathbf{T}^{-1} \qquad \text{(6-40)}$$

Equation (6-40) shows that every square matrix is similar to its transpose.

A special case of a similarity transformation occurs when the transformation matrix **T** is an orthogonal matrix (see Section 4.6 for discussion of orthogonal matrices). If **T** is orthogonal, its inverse is the same as its transpose. Thus, an *orthogonal similarity transformation* (or simply, *orthogonal transformation*) is also a congruence transformation, and it is expressed by the equation

$$\mathbf{B} = \mathbf{T}^{-1}\mathbf{AT} = \mathbf{T}^\mathsf{T}\mathbf{AT} \qquad \text{(6-41)}$$

in which **T** is now an orthogonal matrix. The matrices **A** and **B** are said to be orthogonally similar. Of course, all of the properties of congruent and similar matrices hold for orthogonally similar matrices. Thus, the matrices **A** and **B** in Equation (6-41) have the same rank, the same characteristic equation, the same eigenvalues, the same trace, and their determinants have equal values. Furthermore, if **X** is an eigenvector of **A**, the corresponding eigenvector of **B** is (see Equation 6-39)

$$\mathbf{Y} = \mathbf{T}^{-1}\mathbf{X} = \mathbf{T}^\mathsf{T}\mathbf{X} \qquad \text{(6-42)}$$

An additional property of orthogonally similar matrices is that if **A** is symmetric, then **B** is symmetric also. This property does not apply to similarity transformations in general, however. As in the case of the more general transformations, an orthogonal transformation can also be defined by the equation $\mathbf{B} = \mathbf{TAT}^{-1} = \mathbf{TAT}^{\mathbf{T}}$ if desired.

Similarity transformations were used in the preceding section to diagonalize unsymmetric matrices. In the next section, the diagonalization of symmetric matrices is accomplished by orthogonal similarity transformations.

6.7. Symmetric Matrices. The special case of a symmetric matrix warrants additional consideration, because such matrices are encountered frequently in engineering work and possess certain unique properties. Of course, all of the definitions and properties described in the preceding six sections of this chapter apply to symmetric matrices as well as to unsymmetric ones. Some additional properties pertaining only to symmetric matrices (which include diagonal matrices) are discussed in this section.

Although an unsymmetric matrix may have eigenvalues that are either real or complex, the eigenvalues of a symmetric matrix always are real. However, since this property can be established only by using complex number theory, it will not be proved here. Because the eigenvalues are real, it follows that all of the eigenvectors can be selected as real, which was explained in Section 6.2.

As in the case of unsymmetric matrices, the eigenvectors of a symmetric matrix corresponding to distinct eigenvalues are linearly independent. Furthermore, when an eigenvalue of a symmetric matrix has multiplicity k, there always will be k linearly independent eigenvectors associated with that eigenvalue, which is not necessarily true when the matrix is unsymmetric (see Section 6.3). Therefore, a symmetric matrix of order n always has n linearly independent eigenvectors.

Now consider two different eigenvalues λ_i and λ_j of a symmetric matrix **A**. These eigenvalues satisfy the equations

$$\mathbf{AX}_i = \lambda_i \mathbf{X}_i \tag{a}$$

$$\mathbf{AX}_j = \lambda_j \mathbf{X}_j \tag{b}$$

in which $\mathbf{X}_i$ and $\mathbf{X}_j$ are the eigenvectors associated with λ_i and λ_j, respectively. Both sides of Equation (a) can be premultiplied by $\mathbf{X}_j^{\mathbf{T}}$ (the transpose of the column vector $\mathbf{X}_j$), and both sides of Equation (b) can be premultiplied by $\mathbf{X}_i^{\mathbf{T}}$, to give

$$\mathbf{X}_j^{\mathbf{T}} \mathbf{AX}_i = \lambda_i \mathbf{X}_j^{\mathbf{T}} \mathbf{X}_i \tag{c}$$

$$\mathbf{X}_i^{\mathbf{T}} \mathbf{AX}_j = \lambda_j \mathbf{X}_i^{\mathbf{T}} \mathbf{X}_j \tag{d}$$

Taking the transposes of Equations (c) and (d) yields two more equations:

$$\mathbf{X}_i^{\mathbf{T}} \mathbf{AX}_j = \lambda_i \mathbf{X}_i^{\mathbf{T}} \mathbf{X}_j \tag{e}$$

$$\mathbf{X}_j^{\mathbf{T}} \mathbf{AX}_i = \lambda_j \mathbf{X}_j^{\mathbf{T}} \mathbf{X}_i \tag{f}$$

By observing that the left-hand sides of Equations (c) and (f) are the same, and similarly for Equations (d) and (e), the right-hand sides of these equations may be equated. Thus,

$$\lambda_i \mathbf{X}_j^\mathsf{T} \mathbf{X}_i = \lambda_j \mathbf{X}_j^\mathsf{T} \mathbf{X}_i$$

and

$$\lambda_i \mathbf{X}_i^\mathsf{T} \mathbf{X}_j = \lambda_j \mathbf{X}_i^\mathsf{T} \mathbf{X}_j$$

Rearranging these equations gives

$$(\lambda_i - \lambda_j)\mathbf{X}_j^\mathsf{T} \mathbf{X}_i = (\lambda_i - \lambda_j)\mathbf{X}_i^\mathsf{T} \mathbf{X}_j = 0 \tag{g}$$

Because λ_i and λ_j are assumed to be different eigenvalues, it follows from Equation (g) that

$$\mathbf{X}_j^\mathsf{T} \mathbf{X}_i = \mathbf{X}_i^\mathsf{T} \mathbf{X}_j = 0 \tag{6-43}$$

which shows that the eigenvectors $\mathbf{X}_i$ and $\mathbf{X}_j$ are orthogonal vectors.

The important conclusion has now been obtained that the eigenvectors of a symmetric matrix corresponding to any two different eigenvalues are orthogonal to one another. It follows that any modal matrix **M** constructed for a symmetric matrix with distinct eigenvalues will have columns consisting of mutually orthogonal vectors. These columns are unique except for an arbitrary scalar multiplier of each column. If the eigenvectors are normalized to unit magnitudes, they will form an orthonormal set of vectors. The modal matrix **N** of normalized eigenvectors is then an orthogonal matrix, and its inverse is equal to its transpose. Henceforth, this orthogonal modal matrix will be denoted by the symbol **Q** to distinguish it from **N**, which in general is not orthogonal. The following example illustrates these aspects of a symmetric matrix.

Example 1. Determine the eigenvalues, eigenvectors, and a modal matrix for the following symmetric matrix:

$$\mathbf{A} = \begin{bmatrix} 2 & 2 & 0 \\ 2 & 5 & 0 \\ 0 & 0 & 3 \end{bmatrix} \tag{6-44}$$

The characteristic equation is

$$\begin{vmatrix} 2-\lambda & 2 & 0 \\ 2 & 5-\lambda & 0 \\ 0 & 0 & 3-\lambda \end{vmatrix} = 0$$

or

$$-\lambda^3 + 10\lambda^2 - 27\lambda + 18 = 0$$

The roots of this cubic equation are the three eigenvalues of **A**:

$$\lambda_1 = 6 \qquad \lambda_2 = 3 \qquad \lambda_3 = 1$$

and therefore the spectral matrix is

$$\mathbf{S} = \begin{bmatrix} 6 & 0 & 0 \\ 0 & 3 & 0 \\ 0 & 0 & 1 \end{bmatrix} \tag{6-45}$$

The eigenvectors are found to be

$$\mathbf{X}_1 = \beta_1 \begin{bmatrix} 1 \\ 2 \\ 0 \end{bmatrix} \qquad \mathbf{X}_2 = \beta_2 \begin{bmatrix} 0 \\ 0 \\ 1 \end{bmatrix} \qquad \mathbf{X}_3 = \beta_3 \begin{bmatrix} -2 \\ 1 \\ 0 \end{bmatrix}$$

Note that each pair of eigenvectors is orthogonal and satisfies Equation (6-43). The modal matrix obtained by arbitrarily specifying $\beta_1 = \beta_2 = \beta_3 = 1$ is

$$\mathbf{M} = \begin{bmatrix} 1 & 0 & -2 \\ 2 & 0 & 1 \\ 0 & 1 & 0 \end{bmatrix} \tag{6-46}$$

The order in which the eigenvectors appear in this modal matrix corresponds to the order selected for the eigenvalues in the spectral matrix. The orthogonal modal matrix **Q** is

$$\mathbf{Q} = \begin{bmatrix} \frac{1}{\sqrt{5}} & 0 & -\frac{2}{\sqrt{5}} \\ \frac{2}{\sqrt{5}} & 0 & \frac{1}{\sqrt{5}} \\ 0 & 1 & 0 \end{bmatrix} \tag{6-47}$$

It can easily be verified that **Q** is orthogonal by multiplying it with its transpose $\mathbf{Q}^T$ to obtain the identity matrix.

Now consider a symmetric matrix with repeated eigenvalues. As explained previously, the eigenvectors associated with different eigenvalues are orthogonal; hence, all eigenvectors associated with repeated eigenvalues are orthogonal to any eigenvectors associated with distinct eigenvalues. The eigenvectors corresponding to a repeated eigenvalue are not determined uniquely, because (as explained in Section 6.3) an infinite number of different sets of linearly independent eigenvectors may be found from any given set by forming linear combinations of the given ones. Some of these sets of eigenvectors are orthogonal and some are not. However, it is always possible to obtain an orthogonal set corresponding to the repeated eigenvalues. These eigenvectors are automatically orthogonal to those associated with distinct eigenvalues. Then it is feasible to construct a modal matrix **M** having columns that are mutually orthogonal vectors. By normalizing these columns to unit magnitudes, an orthogonal modal matrix **Q** is obtained.

It is possible for a matrix to have all of its eigenvalues the same, as illustrated in Examples 3, 4, and 5 of Section 6.3. However, for a symmetric matrix, all eigenvalues are equal if and only if the matrix is diagonal. In this case, the diagonal elements themselves are the eigenvalues (see Example 3 of Section 6.3).

Example 2. Determine the eigenvalues, eigenvectors, and a modal matrix for the following symmetric matrix:

$$\mathbf{B} = \begin{bmatrix} 5 & 1 & \sqrt{2} \\ 1 & 5 & \sqrt{2} \\ \sqrt{2} & \sqrt{2} & 6 \end{bmatrix} \tag{6-48}$$

The homogeneous equations obtained from this matrix (see Equation 6-3) are

$$(5-\lambda)X_1 + X_2 + \sqrt{2}X_3 = 0 \tag{6-49a}$$

$$X_1 + (5-\lambda)X_2 + \sqrt{2}X_3 = 0 \tag{6-49b}$$

$$\sqrt{2}X_1 + \sqrt{2}X_2 + (6-\lambda)X_3 = 0 \tag{6-49c}$$

and the characteristic equation is

$$-\lambda^3 + 16\lambda^2 - 80\lambda + 128 = 0$$

Thus, the eigenvalues are

$$\lambda_1 = 8 \qquad \lambda_2 = \lambda_3 = 4 \tag{6-50}$$

and the matrix has a repeated eigenvalue of multiplicity two. The spectral matrix is

$$\mathbf{S} = \begin{bmatrix} 8 & 0 & 0 \\ 0 & 4 & 0 \\ 0 & 0 & 4 \end{bmatrix} \tag{6-51}$$

The eigenvector corresponding to the first eigenvalue is found by substituting $\lambda = 8$ into Equations (6–49) and solving the resulting equations:

$$-3X_1 + X_2 + \sqrt{2}X_3 = 0$$
$$X_1 - 3X_2 + \sqrt{2}X_3 = 0$$
$$\sqrt{2}X_1 + \sqrt{2}X_2 - 2X_3 = 0$$

The rank of these equations is two, hence their solution consists of one linearly independent eigenvector:

$$\mathbf{X}_1 = \beta_1 \begin{bmatrix} 1 \\ 1 \\ \sqrt{2} \end{bmatrix}$$

The eigenvectors corresponding to the repeated eigenvalue are found by solving Equations (6–49) with $\lambda = 4$:

$$X_1 + X_2 + \sqrt{2}X_3 = 0$$
$$X_1 + X_2 + \sqrt{2}X_3 = 0$$
$$\sqrt{2}X_1 + \sqrt{2}X_2 + 2X_3 = 0$$

Because the rank of these equations is one, their solution consists of two linearly independent eigenvectors, as expected for a symmetric matrix. The solution vector for the equations is

$$\begin{bmatrix} -X_2 - \sqrt{2}\,X_3 \\ X_2 \\ X_3 \end{bmatrix}$$

in which X_2 and X_3 are arbitrary. Hence, one set of linearly independent eigenvectors is

$$\mathbf{X}_2 = \beta_2 \begin{bmatrix} -1 \\ 1 \\ 0 \end{bmatrix} \qquad \mathbf{X}_3 = \beta_3 \begin{bmatrix} -\sqrt{2} \\ 0 \\ 1 \end{bmatrix}$$

The first of these vectors is obtained by setting $X_2 = 1$ and $X_3 = 0$, the second by setting $X_2 = 0$ and $X_3 = 1$. Note that the first eigenvector is orthogonal to $\mathbf{X}_2$ and $\mathbf{X}_3$, as expected, but $\mathbf{X}_2$ and $\mathbf{X}_3$ are not orthogonal to each other. An infinite number of sets of linearly independent eigenvectors $\mathbf{X}_2$ and $\mathbf{X}_3$ can be constructed by assigning various values to X_2 and X_3 in the solution vector or by forming linear combinations of the eigenvectors $\mathbf{X}_2$ and $\mathbf{X}_3$.

The eigenvectors $\mathbf{X}_2$ and $\mathbf{X}_3$ can be selected as orthogonal vectors if desired, and again an infinite number of pairs of eigenvectors exist. It is relatively simple to construct orthogonal eigenvectors for this case. The procedure is to assume $\mathbf{X}_2$ and $\mathbf{X}_3$ as linear combinations of the arbitrary vectors selected previously:

$$\mathbf{X}_2 = \alpha_1 \begin{bmatrix} -1 \\ 1 \\ 0 \end{bmatrix} + \alpha_2 \begin{bmatrix} -\sqrt{2} \\ 0 \\ 1 \end{bmatrix}$$

$$\mathbf{X}_3 = \alpha_3 \begin{bmatrix} -1 \\ 1 \\ 0 \end{bmatrix} + \alpha_4 \begin{bmatrix} -\sqrt{2} \\ 0 \\ 1 \end{bmatrix}$$

and then to use the orthogonality condition $\mathbf{X}_2^{\mathbf{T}}\mathbf{X}_3 = 0$ to determine the constants. However, it is just as general, and the algebra is made easier, if α_1 and α_3 are taken as unity. With this simplification, the orthogonality condition yields the equation

$$(-1 - \sqrt{2}\alpha_2)(-1 - \sqrt{2}\alpha_4) + 1 + \alpha_2\alpha_4 = 0$$

or

$$\sqrt{2}(\alpha_2 + \alpha_4) + 3\alpha_2\alpha_4 + 2 = 0$$

An infinite number of solutions of this equation exists, and each solution produces a pair of orthogonal eigenvectors. As an example, suppose that $\alpha_2 = 0$ and $\alpha_4 = -\sqrt{2}$; then the eigenvectors become

$$\mathbf{X}_2 = \beta_2 \begin{bmatrix} -1 \\ 1 \\ 0 \end{bmatrix} \qquad \mathbf{X}_3 = \beta_3 \begin{bmatrix} 1 \\ 1 \\ -\sqrt{2} \end{bmatrix}$$

Since these vectors are orthogonal to one another, all three eigenvectors are orthogonal. Therefore, the corresponding modal matrix $\mathbf{M}$ contains columns that are not only linearly independent but also orthogonal. When these columns are normalized to unit magnitudes, an orthogonal modal matrix $\mathbf{Q}$ is obtained:

$$\mathbf{Q} = \begin{bmatrix} \frac{1}{2} & -\frac{1}{\sqrt{2}} & \frac{1}{2} \\ \frac{1}{2} & \frac{1}{\sqrt{2}} & \frac{1}{2} \\ \frac{\sqrt{2}}{2} & 0 & -\frac{\sqrt{2}}{2} \end{bmatrix}$$

Of course, the product of $\mathbf{Q}$ and its transpose $\mathbf{Q}^{\mathbf{T}}$ is the identity matrix.

The procedure illustrated in the preceding example for determining orthogonal eigenvectors corresponding to a repeated eigenvalue is relatively simple to use. In each case, it is necessary to begin by arbitrarily selecting a set of linearly independent eigenvectors. Then new eigenvectors are expressed as linear combinations of these arbitrary ones; these combinations involve unknown constants α (see Equation 5-15). The orthogonality conditions among the new eigenvectors produce equations that must be satisfied by the constants α. When the α's are selected so as to satisfy these equations, the result will be a set of orthogonal eigenvectors.*

Every symmetric matrix **A** of order n can be diagonalized by the similarity transformation shown in Equation (6-32), because a symmetric matrix has n linearly independent eigenvectors. If the orthogonal matrix **Q** is used for this purpose, Equation (6-32) can be rewritten in the following special form:

$$\mathbf{Q}^{-1}\mathbf{A}\mathbf{Q} = \mathbf{Q}^{\mathrm{T}}\mathbf{A}\mathbf{Q} = \mathbf{S} \tag{6-52}$$

This equation represents an orthogonal similarity transformation (see Equation 6-41) on the matrix **A** and shows that every symmetric matrix is orthogonally similar to a diagonal matrix. Equation (6-52) can easily be verified for the matrices **A** and **B** discussed in Examples 1 and 2.

Although they do not serve an immediate purpose, some final definitions pertaining to a symmetric matrix will be mentioned here for the sake of completeness. A symmetric matrix is said to be *positive definite* if all of its eigenvalues are positive; if all are negative, the matrix is said to be *negative definite*. If some eigenvalues are positive and some are negative, the matrix is *indefinite*. Finally, if all eigenvalues are positive and at least one is zero, the matrix is *positive semi-definite*; and if all are negative and at least one is zero, it is *negative semi-definite*. The definiteness of a matrix is not altered either by a congruence transformation or by a similarity transformation.

6.8. Square Roots and Powers of a Matrix. The square root of a square matrix was defined previously in Section 2.5 (see Equations 2-31 and 2-32), and a few simple examples of square roots were given. However, a general method for calculating square roots requires the use of eigenvalues and eigenvectors. To develop the method, consider again Equation (6-31), which holds for any square matrix **A** of order n having n linearly independent eigenvectors. The existence of such independent eigenvectors is assured if the matrix **A** is symmetric or if it is unsymmetric and has n distinct eigenvalues. In addition, some unsymmetric matrices with repeated eigenvalues also have n linearly independent eigenvectors (see Section 6.3). Postmultiplying both sides of Equation (6-31) by the inverse of the modal matrix **M** yields

$$\mathbf{A} = \mathbf{M}\mathbf{S}\mathbf{M}^{-1} \tag{6-53}$$

---

*A formal mathematical procedure for constructing an orthogonal set of eigenvectors from a given set of nonorthogonal eigenvectors is the Gram-Schmidt orthogonalization process (see References 1 and 3).

Now consider a square matrix **B** of order n, given as follows:

$$\mathbf{B} = \mathbf{M}\sqrt{\mathbf{S}}\,\mathbf{M}^{-1}$$

in which $\sqrt{\mathbf{S}}$ represents a square root of the diagonal spectral matrix **S**. (For a discussion of square roots of a diagonal matrix, see Section 2.5.) Taking the square of **B** gives

$$\mathbf{BB} = \mathbf{B}^2 = \mathbf{M}\sqrt{\mathbf{S}}\,\mathbf{M}^{-1}\mathbf{M}\sqrt{\mathbf{S}}\,\mathbf{M}^{-1} = \mathbf{MSM}^{-1} = \mathbf{A}$$

Hence, the matrix **B** is a square root of **A**, or

$$\mathbf{B} = \sqrt{\mathbf{A}} = \mathbf{M}\sqrt{\mathbf{S}}\,\mathbf{M}^{-1} \tag{6-54}$$

In the special case where **A** is a symmetric matrix, the eigenvectors can be normalized, and the modal matrix **M** becomes an orthogonal matrix **Q**. Then,

$$\mathbf{B} = \sqrt{\mathbf{A}} = \mathbf{Q}\sqrt{\mathbf{S}}\,\mathbf{Q}^{\mathrm{T}} \tag{6-55}$$

The use of an orthogonal modal matrix eliminates the need to obtain a matrix inverse.

The procedure for finding a square root of a square matrix **A** may now be summarized. First, obtain eigenvalues and eigenvectors of **A** and assemble them into a spectral matrix **S** and a modal matrix **M** (or normalized modal matrix **Q**). Of course, the order in which the eigenvectors are placed as columns of **M** must agree with the order of the eigenvalues in **S**. Second, obtain a square root of **S** by taking square roots of the individual eigenvalues appearing on the diagonal. Finally, calculate a square root of **A** from Equation (6–54) or (6–55). By using different square roots of **S**, different square roots of **A** are obtained. However, changing the scale factors for any of the eigenvectors in **M** will have no effect on the results.

In general, there are 2^n different square roots of the spectral matrix **S** because there are two values (positive and negative) for the square root of each eigenvalue. Thus, an nth order square matrix normally has 2^n different square roots. There are exceptions, however, such as the matrices described in Section 2.5 which have an infinite number of square roots and the 2×2 matrix of Example 2 which follows. This latter matrix has a zero eigenvalue, so there are only two square roots (instead of four) of the spectral matrix.

If all eigenvalues of a matrix **A** are real and positive, then their square roots will be real; hence, the matrix square root also will be real. If all eigenvalues are real but some are negative, then the square roots of the negative eigenvalues are imaginary. In this case the matrix square root may contain complex terms. Although the method described in this section for finding square roots is suitable for matrices with real eigenvalues that are either positive or negative, the following examples are limited to the former case (which means that the square roots are real).

Example 1. Determine the square roots of the symmetric matrix **A**:

$$\mathbf{A} = \begin{bmatrix} 5 & -2 \\ -2 & 8 \end{bmatrix}$$

The eigenvalues of this matrix, found by the method of Section 6.1, are

$$\lambda_1 = 9 \qquad \lambda_2 = 4$$

and, therefore, the spectral matrix and its four square roots are:

$$\mathbf{S} = \begin{bmatrix} 9 & 0 \\ 0 & 4 \end{bmatrix}$$

$$\sqrt{\mathbf{S}} = \begin{bmatrix} 3 & 0 \\ 0 & 2 \end{bmatrix} \qquad \sqrt{\mathbf{S}} = \begin{bmatrix} 3 & 0 \\ 0 & -2 \end{bmatrix}$$

$$\sqrt{\mathbf{S}} = \begin{bmatrix} -3 & 0 \\ 0 & 2 \end{bmatrix} \qquad \sqrt{\mathbf{S}} = \begin{bmatrix} -3 & 0 \\ 0 & -2 \end{bmatrix}$$

Next, the eigenvectors are determined and a modal matrix (and its inverse) obtained:

$$\mathbf{M} = \begin{bmatrix} -1 & 2 \\ 2 & 1 \end{bmatrix} \qquad \mathbf{M}^{-1} = \frac{1}{5}\begin{bmatrix} -1 & 2 \\ 2 & 1 \end{bmatrix}$$

Four square roots of **A** can be calculated by using Equation (6-54) with the four square roots of **S**. These roots are:

$$\sqrt{\mathbf{A}} = \frac{1}{5}\begin{bmatrix} 11 & -2 \\ -2 & 14 \end{bmatrix} \qquad \sqrt{\mathbf{A}} = \begin{bmatrix} -1 & -2 \\ -2 & 2 \end{bmatrix}$$

$$\sqrt{\mathbf{A}} = \begin{bmatrix} 1 & 2 \\ 2 & -2 \end{bmatrix} \qquad \sqrt{\mathbf{A}} = \frac{1}{5}\begin{bmatrix} -11 & 2 \\ 2 & -14 \end{bmatrix}$$

Because two of the square roots of **S** are the negatives of the other two, it is apparent that two of the square roots of **A** are also negatives of the others. Hence, it is only necessary to perform the multiplications indicated by Equation (6-54) twice instead of four times:

Example 2. Determine the square roots of the following unsymmetric matrix:

$$\mathbf{A} = \begin{bmatrix} 4 & -2 \\ 6 & -3 \end{bmatrix}$$

The eigenvalues of this matrix are 1 and 0, giving a spectral matrix with only two square roots:

$$\mathbf{S} = \begin{bmatrix} 1 & 0 \\ 0 & 0 \end{bmatrix} \qquad \sqrt{\mathbf{S}} = \begin{bmatrix} 1 & 0 \\ 0 & 0 \end{bmatrix} \qquad \sqrt{\mathbf{S}} = \begin{bmatrix} -1 & 0 \\ 0 & 0 \end{bmatrix}$$

A modal matrix and its inverse are

$$\mathbf{M} = \begin{bmatrix} 2 & 1 \\ 3 & 2 \end{bmatrix} \qquad \mathbf{M}^{-1} = \begin{bmatrix} 2 & -1 \\ -3 & 2 \end{bmatrix}$$

A square root of **A** can now be obtained from Equation (6-54) by using the first square root of **S**:

$$\sqrt{\mathbf{A}} = \begin{bmatrix} 4 & -2 \\ 6 & -3 \end{bmatrix}$$

Thus, it happens that the matrix **A** is its own square root, showing that this relationship is not limited to the identity matrix. Of course, the second square root of **A** is the negative of the one shown above.

Example 3. Determine the square roots of the symmetric matrix **A**:

$$\mathbf{A} = \begin{bmatrix} 5 & 8 & 2 \\ 8 & 14 & 5 \\ 2 & 5 & 5 \end{bmatrix}$$

The characteristic equation is

$$\lambda^3 - 24\lambda^2 + 72\lambda - 9 = 0$$

and hence the eigenvalues (found by trial and error) are

$$\lambda_1 = 20.511 \qquad \lambda_2 = 3.358 \qquad \lambda_3 = 0.1307$$

The spectral matrix and its square roots become

$$\mathbf{S} = \begin{bmatrix} 20.511 & 0 & 0 \\ 0 & 3.358 & 0 \\ 0 & 0 & 0.1307 \end{bmatrix} \quad \sqrt{\mathbf{S}} = \begin{bmatrix} \pm 4.529 & 0 & 0 \\ 0 & \pm 1.833 & 0 \\ 0 & 0 & \pm 0.3615 \end{bmatrix}$$

and the eigenvectors are

$$\mathbf{X}_1 = \beta_1 \begin{bmatrix} 1.433 \\ 2.529 \\ 1 \end{bmatrix} \qquad \mathbf{X}_2 = \beta_2 \begin{bmatrix} -0.4023 \\ -0.1675 \\ 1 \end{bmatrix} \qquad \mathbf{X}_3 = \beta_3 \begin{bmatrix} 3.469 \\ -2.361 \\ 1 \end{bmatrix}$$

When these vectors are normalized to unit magnitudes, the orthogonal modal matrix **Q** is obtained:

$$\mathbf{Q} = \begin{bmatrix} 0.4662 & -0.3688 & 0.8041 \\ 0.8227 & -0.1535 & -0.5474 \\ 0.3253 & 0.9168 & 0.2318 \end{bmatrix}$$

Finally, the square roots are found by substituting $\sqrt{\mathbf{S}}$ and **Q** into Equation (6-55) and multiplying. If the square root of **S** having all positive terms is used, the following square root of **A** is obtained:

$$\sqrt{\mathbf{A}} = \begin{bmatrix} 1.467 & 1.682 & 0.135 \\ 1.682 & 3.217 & 0.908 \\ 0.135 & 0.908 & 2.039 \end{bmatrix}$$

If the square root of **S** having the first two diagonal terms positive and the last one negative is used, the result is

$$\sqrt{\mathbf{A}} = \begin{bmatrix} 1 & 2 & 0 \\ 2 & 3 & 1 \\ 0 & 1 & 2 \end{bmatrix}$$

Two additional square roots can be found by using other forms of $\sqrt{\mathbf{S}}$, and then four more can be found simply by taking the negatives of the preceding ones. Thus, there are eight square roots of the 3×3 matrix **A**.

Powers of a Matrix. Positive integer powers of a matrix also can be calculated by using the modal matrix **M** and the spectral matrix **S**. To illustrate the concept, consider the calculation of the square of a matrix **A** by utilizing the relationship given in Equation (6-53):

$$\mathbf{A}^2 = \mathbf{MSM}^{-1}\mathbf{MSM}^{-1} = \mathbf{MS}^2\mathbf{M}^{-1}$$

This process can be repeated for higher integer powers, yielding the general result

$$\mathbf{A}^m = \mathbf{MS}^m\mathbf{M}^{-1} \tag{6-56}$$

In this formula, **A** is a square matrix of order n having n linearly independent eigenvectors, **M** is a modal matrix of eigenvectors, and **S** is the diagonal spectral matrix. Integer powers of **S** are simple to calculate, because each term on the diagonal is raised to the mth power. When the matrix **A** is symmetric, the eigenvectors can be normalized; then Equation (6-56) becomes

$$\mathbf{A}^m = \mathbf{QS}^m\mathbf{Q}^{\mathrm{T}} \tag{6-57}$$

in which **Q** is an orthogonal matrix.

The eigenvalues of the power of a matrix can be determined readily from the eigenvalues of the original matrix. To obtain the pertinent relationship, multiply both sides of Equation (6-1) by **A**:

$$\mathbf{A}^2\mathbf{X} = \lambda\mathbf{A}\mathbf{X}$$

Since $\mathbf{AX} = \lambda\mathbf{X}$, this equation becomes

$$\mathbf{A}^2\mathbf{X} = \lambda^2\mathbf{X}$$

Continuing in this manner yields the general equation

$$\mathbf{A}^m\mathbf{X} = \lambda^m\mathbf{X} \tag{6-58}$$

which shows that λ^m is an eigenvalue of $\mathbf{A}^m$. It also shows that the matrix $\mathbf{A}^m$ has the same eigenvectors as **A**. Of course, the power m must be a positive integer.

Negative integer powers may be handled in the following manner:

$$\mathbf{A}^{-m} = (\mathbf{A}^m)^{-1} = (\mathbf{M}\mathbf{S}^m\mathbf{M}^{-1})^{-1} = \mathbf{M}(\mathbf{S}^m)^{-1}\mathbf{M}^{-1}$$

or

$$\mathbf{A}^{-m} = \mathbf{M}\mathbf{S}^{-m}\mathbf{M}^{-1} \tag{6-59}$$

The matrix $\mathbf{S}^{-m}$, which is the inverse of $\mathbf{S}^m$, is easily found from $\mathbf{S}^m$ by taking the reciprocals of the diagonal terms. For a symmetric matrix the formula becomes

$$\mathbf{A}^{-m} = \mathbf{Q}\mathbf{S}^{-m}\mathbf{Q}^{\mathsf{T}} \tag{6-60}$$

When m has a high value, the preceding formulas for positive and negative integer powers may be easier to use than direct multiplication.

6.9. Method of Iteration for Eigenvalues and Eigenvectors. The determination of eigenvalues and eigenvectors for a matrix of order greater than 2×2 may require considerable calculating effort if the work is done by hand. The fundamental procedures described in previous sections are only suitable for small matrices or for those having convenient numerical values, as was the case in the examples of those sections. Hence, many numerical methods have been developed for determining eigenvalues and eigenvectors. The method presented in this section is suitable for practical use with matrices of order 3×3 or 4×4; for larger matrices the use of a digital computer is almost a necessity. The method to be described is an iterative procedure known as the Stodola-Vianello method.* It is more suitable for symmetric matrices than for unsymmetric matrices, although it can be used for either kind. The meth-

*The method was suggested first by the Italian engineer Luigi Vianello (1862–1907), who presented it in 1898 for calculating buckling loads of columns; later, the Swiss engineer Aurel Stodola (1859–1942) applied the method to vibrations of shafts. For information on other numerical methods, references on numerical analysis should be consulted.

od requires that a good approximation to the eigenvectors be known in advance; otherwise, the convergence of the method is too slow for practical use. In spite of this difficulty, the method has found wide application in engineering problems because usually the general "shape" of the eigenvector is known from the physical behavior (for example, the eigenvector may represent the mode shape of a vibrating element). In such cases, the trial eigenvector can be estimated with a fair degree of accuracy, and the convergence of the iteration procedure will be sufficiently rapid.

In order to begin the development of the iteration method, refer again to the matrix equation of the eigenvalue problem (Equation 6–1). A first approximation to one of the eigenvalues λ may be obtained by substituting a trial eigenvector $\mathbf{X}_{(1)}$ into both sides of the equation and then solving for λ. Let the product of the matrix $\mathbf{A}$ and the trial eigenvector $\mathbf{X}_{(1)}$ on the left-hand side of the equation be represented by the vector $\mathbf{Y}_{(1)}$; thus,

$$\mathbf{A}\mathbf{X}_{(1)} = \mathbf{Y}_{(1)} \tag{a}$$

If the vector $\mathbf{X}_{(1)}$ is not the true eigenvector, then its substitution into Equation (6–1) will only satisfy the equation approximately. Therefore,

$$\mathbf{Y}_{(1)} \approx \lambda \mathbf{X}_{(1)}$$

A first approximation $\lambda_{(1)}$ to the eigenvalue λ may be obtained by dividing any element of the vector $\mathbf{Y}_{(1)}$ by the corresponding element of the vector $\mathbf{X}_{(1)}$. Of course, if $\mathbf{X}_{(1)}$ were the true eigenvector, all such ratios would be equal. Therefore, several possibilities exist for the determination of $\lambda_{(1)}$. As an arbitrary procedure, a particular element (such as the first or last one) can always be selected for determining the ratio. However, better results are obtained by taking the ratio of the sums of all the elements in the vectors. In other words, the elements of the vector $\mathbf{Y}_{(1)}$ are summed, the elements of the vector $\mathbf{X}_{(1)}$ are summed, and then the ratio of these two sums is taken as the first approximation $\lambda_{(1)}$:

$$\lambda_{(1)} = \frac{\text{Sum of elements of } \mathbf{Y}_{(1)}}{\text{Sum of elements of } \mathbf{X}_{(1)}} \tag{b}$$

The value given in Equation (b) is always intermediate between the largest and smallest ratios obtained by taking ratios of particular elements.

A second step of iteration can now be performed by taking the vector $\mathbf{Y}_{(1)}$ (see Equation a) as the second trial eigenvector. At this stage of the procedure the new trial eigenvector can be scaled (as a matter of convenience) to avoid unusually large or small numbers. That is, an arbitrary constant $\beta_{(1)}$ may be used as a scale factor on $\mathbf{Y}_{(1)}$ to produce the new trial eigenvector $\mathbf{X}_{(2)}$:

$$\mathbf{X}_{(2)} = \beta_{(1)} \mathbf{Y}_{(1)} \tag{c}$$

The vector $\mathbf{X}_{(2)}$ is then premultiplied by $\mathbf{A}$ to give

$$\mathbf{AX}_{(2)} = \mathbf{Y}_{(2)}$$

Next, a second approximation $\lambda_{(2)}$ is found by taking the ratio of the sum of the elements in $\mathbf{Y}_{(2)}$ to the sum of the elements in $\mathbf{X}_{(2)}$. This procedure is repeated as many times as necessary until the eigenvalue and the associated eigenvector are determined to the desired degree of accuracy. At each stage of the iteration procedure the new trial eigenvector can be scaled if desired.

The iteration procedure described above can be summarized by the following equations for the kth iteration:

$$\mathbf{AX}_{(k)} = \mathbf{Y}_{(k)} \tag{6-61a}$$

$$\lambda_{(k)} = \frac{\text{Sum of elements of } \mathbf{Y}_{(k)}}{\text{Sum of elements of } \mathbf{X}_{(k)}} \tag{6-61b}$$

$$\mathbf{X}_{(k+1)} = \beta_{(k)}\mathbf{Y}_{(k)} \tag{6-61c}$$

in which $\beta_{(k)}$ is an arbitrarily selected scaling factor. It can be shown that when these equations are applied repeatedly, the procedure will converge to the numerically largest eigenvalue λ_1 and the associated eigenvector $\mathbf{X}_1$ of the matrix $\mathbf{A}$. Therefore, for the purpose of aiding convergence, the first trial eigenvector (see Equation a) must be an approximation to the eigenvector associated with the numerically largest eigenvalue.

In order to determine the second largest eigenvalue λ_2 and the corresponding eigenvector $\mathbf{X}_2$, the order of the matrix $\mathbf{A}$ must be reduced by an elimination process. This process uses the orthogonality property of the eigenvectors when the matrix is symmetric, as described in the next paragraph; for unsymmetric matrices it is necessary to consider the eigenvectors of the transpose of the original matrix, and this procedure is described at the end of the section. The iteration procedure is applied to the new matrix of order $n - 1$ to obtain λ_2 and $\mathbf{X}_2$. This process of matrix reduction followed by iteration is continued until all eigenvalues and eigenvectors are determined. When the matrix has been reduced to a 2×2 matrix, the remaining eigenvalues can be found either by iteration or by solving the characteristic equation (which is quadratic).

The process of matrix reduction can be accomplished in the following way. The orthogonality condition between the first eigenvector $\mathbf{X}_1$ (which is known) and any other eigenvector $\mathbf{X}_i$ is

$$\mathbf{X}_1^{\mathbf{T}}\mathbf{X}_i = 0$$

or, when the product is formed,

$$aX_1 + bX_2 + cX_3 + \cdots = 0 \tag{d}$$

in which $a, b, c, \ldots$ are the known elements of $\mathbf{X}_1$ and $X_1, X_2, X_3, \ldots$ are the unknown elements of $\mathbf{X}_i$. Equation (d) can be solved for X_1 in terms of $X_2, X_3, \ldots$; thus,

$$X_1 = -\frac{b}{a}X_2 - \frac{c}{a}X_3 - \cdots \tag{e}$$

Then, this expression for X_1 can be substituted into the original equations of the eigenvalue problem (see Equation 6-1) to produce n equations in $n - 1$ unknowns. The first of these equations will be a linear combination of the remaining $n - 1$ equations and can be discarded, leaving the reduced set of $n - 1$ equations in $n - 1$ unknowns:

$$\mathbf{BX} = \lambda\mathbf{X} \tag{f}$$

Equation (f) constitutes an eigenvalue problem involving only the $n - 1$ remaining eigenvalues of the original system of equations. Now the entire process of iteration can be repeated for the new matrix $\mathbf{B}$ of order $n - 1$. The result will be the numerically highest eigenvalue of the reduced matrix (which is the second eigenvalue of the original matrix $\mathbf{A}$) and the corresponding eigenvector for the reduced matrix. This reduced eigenvector contains only the values of $X_2, X_3, \ldots, X_n$ of the eigenvector for the original matrix. The associated value of X_1 can be found from Equation (e). Thus, the second eigenvector of the original matrix is determined. The iteration and reduction process can be continued until a 2×2 matrix is obtained, after which the last two eigenvalues and eigenvectors can be found by solving the characteristic equation. The following illustrative example will serve to demonstrate the method.

Example. Calculate the eigenvalues and eigenvectors for the following symmetric matrix $\mathbf{A}$ using the iteration procedure:

$$\mathbf{A} = \begin{bmatrix} 4.83 & 7.29 & 2.06 \\ 7.29 & 5.91 & 3.42 \\ 2.06 & 3.42 & 2.96 \end{bmatrix} \tag{6-62}$$

Assume that the numerically largest eigenvalue and its associated eigenvector are known to have the following approximate values:

$$\lambda_1 \approx 15 \qquad \mathbf{X}_1 \approx \beta_1 \{1 \quad 1 \quad 0.5\}$$

The approximate eigenvector $\mathbf{X}_1$ will be used as the first trial vector $\mathbf{X}_{(1)}$:

$$\mathbf{X}_{(1)} = \{1 \quad 1 \quad 0.5\}$$

Premultiplication of $\mathbf{X}_{(1)}$ by $\mathbf{A}$ produces the vector

$$\mathbf{Y}_{(1)} = \{13.15 \quad 14.91 \quad 6.96\}$$

Taking the sum of all elements in $\mathbf{Y}_{(1)}$, and dividing by the sum of all elements in $\mathbf{X}_{(1)}$, gives the first approximation to the eigenvalue:

$$\lambda_{(1)} = \frac{35.02}{2.5} = 14.01$$

The results of this first cycle of the solution are given in Columns 1 and 2 of Table 6-1.

The second step in the iteration procedure begins by taking $\mathbf{Y}_{(1)}$ as the second trial eigenvector $\mathbf{X}_{(2)}$. The vector can be taken exactly as expressed in Column 2 of Table 6-1, or it can be scaled. For convenience in this example, all trial eigenvectors are scaled so that the first element is equal to unity. Therefore, the second trial vector $\mathbf{X}_2$ is taken as $\mathbf{Y}_{(1)}$ divided by 13.15, which is the value of the first element of $\mathbf{Y}_{(1)}$. The vector $\mathbf{X}_{(2)}$ is recorded in Column 3 of the table, and the result of premultiplying it by the matrix $\mathbf{A}$ is

Table 6-1

Example of Iteration Method

$$\mathbf{A} = \begin{bmatrix} 4.83 & 7.29 & 2.06 \\ 7.29 & 5.91 & 3.42 \\ 2.06 & 3.42 & 2.96 \end{bmatrix}$$

Cycle	1		2		3		4	
Column Number	1	2	3	4	5	6	7	8
Trial Vector	$\mathbf{X}_{(1)}$	$\mathbf{AX}_{(1)} = \mathbf{Y}_{(1)}$	$\mathbf{X}_{(2)}$	$\mathbf{AX}_{(2)} = \mathbf{Y}_{(2)}$	$\mathbf{X}_{(3)}$	$\mathbf{AX}_{(3)} = \mathbf{Y}_{(3)}$	$\mathbf{X}_{(4)}$	$\mathbf{AX}_{(4)} = \mathbf{Y}_{(4)}$
X_1	1.00	13.15	1.00	14.16	1.00	14.01	1.00	14.09
X_2	1.00	14.91	1.13	15.78	1.11	15.66	1.12	15.72
X_3	0.50	6.96	0.53	7.49	0.53	7.42	0.53	7.46
Sum	2.50	35.02	2.66	37.43	2.64	37.09	2.65	37.27
λ	14.01		14.07		14.05		14.06	

recorded in Column 4. Now a second approximation to the eigenvalue can be obtained as shown in the two lines at the bottom of the table:

$$\lambda_{(2)} = \frac{37.43}{2.66} = 14.07$$

The iteration procedure is continued through additional cycles as shown in Table 6-1 until two successive cycles produce results that agree within a chosen degree of accuracy. Thus, after four cycles the following values are obtained:

$$\lambda_1 = 14.06 \qquad \mathbf{X}_1 = \beta_1 \{1.00 \quad 1.12 \quad 0.53\} \tag{g}$$

The next step in the calculations consists of reducing the order of the matrix **A** by one, using the orthogonality conditions denoted by Equation (d):

$$1.00X_1 + 1.12X_2 + 0.53X_3 = 0$$

Solving for X_1 as in Equation (e),

$$X_1 = -1.12X_2 - 0.53X_3 \tag{h}$$

Substitution of X_1 from Equation (h) into all but the first of the original equations of the eigenvalue problem (see Equation 6-1) produces two equations in two unknowns, as follows:

$$\begin{aligned} -2.25X_2 - 0.44X_3 &= \lambda X_2 \\ 1.11X_2 + 1.87X_3 &= \lambda X_3 \end{aligned} \tag{i}$$

If the value of X_1 from Equation (h) is substituted into the first of the original equations, it will be found that the resulting equation is a linear combination of the other two and can be discarded. Thus, the reduced set of equations is

$$\begin{bmatrix} -2.25 & -0.44 \\ 1.11 & 1.87 \end{bmatrix} \begin{bmatrix} X_2 \\ X_3 \end{bmatrix} = \lambda \begin{bmatrix} X_2 \\ X_3 \end{bmatrix} \tag{j}$$

which constitutes an eigenvalue problem involving only the second and third eigenvalues of the original system. Comparison of Equation (j) with Equation (f) shows that the reduced matrix of order 2 that remains is

$$\mathbf{B} = \begin{bmatrix} -2.25 & -0.44 \\ 1.11 & 1.87 \end{bmatrix}$$

The eigenvalues and eigenvectors of the matrix **B** can be determined by solving the characteristic equation obtained from **B**:

$$(-2.25-\lambda)(1.87-\lambda) + 0.44(1.11) = 0$$

or

$$\lambda^2 + 0.38\lambda - 3.72 = 0$$

The two roots of this quadratic equation are

$$\lambda_2 = -2.13 \qquad \lambda_3 = 1.75 \tag{k}$$

and these quantities are the remaining two eigenvalues of the matrix **A**. The eigenvectors of **B** corresponding to λ_2 and λ_3 are

$$\mathbf{X}_2 = \beta_2 \{1.0 \quad -0.28\}$$

$$\mathbf{X}_3 = \beta_3 \{1.0 \quad -9.1\}$$

These vectors contain only the last two elements of the eigenvectors of **A**; the first element in each case can be found by applying Equation (h). Thus, for the second eigenvector,

$$X_1 = -1.12(1.0) - 0.53(-0.28) = -0.97$$

and for the third eigenvector,

$$X_1 = -1.12(1.0) - 0.53(-9.1) = 3.7$$

Therefore, the elements of the second and third eigenvectors of **A** are, respectively,

$$\{-0.97 \quad 1.0 \quad -0.28\}$$

$$\{3.7 \quad 1.0 \quad -9.1\}$$

or, when scaled so that the first element is unity,

$$\mathbf{X}_2 = \beta_2 \{1.0 \quad -1.03 \quad 0.29\} \tag{l}$$

$$\mathbf{X}_3 = \beta_3 \{1.0 \quad 0.27 \quad -2.5\} \tag{m}$$

Thus, the eigenvalues and eigenvectors of the matrix **A** are given by Equations (g), (k), (l), and (m). More precise results can be obtained by extending Table 6-1 to additional cycles of iteration and by using more significant digits in the numerical calculations. The following results, accurate to four significant figures, can be compared with those found above:

$$\begin{aligned} \lambda_1 &= 14.06 & \mathbf{X}_1 &= \beta_1 \{1.000 \quad 1.117 \quad 0.5296\} \\ \lambda_2 &= -2.106 & \mathbf{X}_2 &= \beta_2 \{1.000 \quad -1.034 \quad 0.2912\} \\ \lambda_3 &= 1.745 & \mathbf{X}_3 &= \beta_3 \{1.000 \quad 0.2731 \quad -2.464\} \end{aligned} \tag{6-63}$$

The iteration method can be used for unsymmetric matrices, as mentioned at the beginning of the section. However, it becomes necessary to work with both the unsymmetric matrix **A** and its transpose $\mathbf{A}^T$. The eigenvalues for **A** and $\mathbf{A}^T$ are the same, and the eigenvectors corresponding to different eigenvalues are orthogonal, as explained in Sections 6.4 and 6.5. The iteration method can be used to obtain the numerically largest eigenvalue and associated eigenvectors for both **A** and $\mathbf{A}^T$ in parallel. Since the first eigenvector of $\mathbf{A}^T$ is orthogonal to the remaining unknown eigenvectors of **A**, its orthogonality property can be used to accomplish the reduction of the matrix **A** to a matrix of lower order. Similarly, the first eigenvector of **A** can be used to accomplish the reduction of $\mathbf{A}^T$ to a matrix of lower order. Thus, by working with both matrices simultaneously there will always be a sufficient number of orthogonality conditions to accomplish the necessary reductions in the order of the matrices. Otherwise, the iteration procedure is the same as for symmetric matrices.

Problems

6.1-1. Determine the characteristic polynomials for each of the following matrices:

$$\mathbf{A} = \begin{bmatrix} 0 & 1 & 0 \\ 0 & 0 & 1 \\ 1 & -3 & 3 \end{bmatrix} \qquad \mathbf{B} = \begin{bmatrix} 1 & 1 & 0 \\ 0 & 1 & 1 \\ 0 & 0 & 1 \end{bmatrix}$$

6.1-2. Show that the following nth order upper triangular matrix has the characteristic polynomial $f(\lambda) = (a - \lambda)^n$.

$$\mathbf{C} = \begin{bmatrix} a & b_1 & b_2 & \cdots & b_{n-2} & b_{n-1} \\ 0 & a & c_1 & \cdots & c_{n-3} & c_{n-2} \\ \cdots & \cdots & \cdots & \cdots & \cdots & \cdots \\ 0 & 0 & 0 & \cdots & a & x \\ 0 & 0 & 0 & \cdots & 0 & a \end{bmatrix}$$

6.1-3. Calculate the eigenvalues of the following symmetric matrix:

$$\mathbf{D} = \begin{bmatrix} 53 & 18 \\ 18 & 38 \end{bmatrix}$$

6.1-4. Calculate the eigenvalues of the following matrix:

$$\mathbf{E} = \begin{bmatrix} 7 & -18 & 3 \\ 0 & -2 & 0 \\ 3 & -6 & 7 \end{bmatrix}$$

6.2-1. Find the eigenvalues and a set of corresponding eigenvectors for the matrix **A** shown below:

$$\mathbf{A} = \begin{bmatrix} -8 & -10 \\ 15 & 17 \end{bmatrix}$$

6.2–2. Solve the preceding problem for the matrix **B**:

$$\mathbf{B} = \begin{bmatrix} 12 & -5 \\ 10 & -3 \end{bmatrix}$$

6.2–3. Solve Problem 6.2–1 for the matrix **C**:

$$\mathbf{C} = \begin{bmatrix} 101 & -245 \\ 42 & -102 \end{bmatrix}$$

6.2–4. Solve Problem 6.2–1 for the diagonal matrix **D**:

$$\mathbf{D} = \begin{bmatrix} 13 & 0 \\ 0 & 2 \end{bmatrix}$$

6.2–5. Solve Problem 6.2–1 for the following matrix:

$$\mathbf{E} = \begin{bmatrix} -20 & 20 \\ -30 & 30 \end{bmatrix}$$

6.2–6. Solve Problem 6.2–1 for the matrix **F**:

$$\mathbf{F} = \begin{bmatrix} 2 & -2 & 0 \\ 0 & 4 & 0 \\ 2 & 5 & 1 \end{bmatrix}$$

6.2–7. Solve Problem 6.2–1 for the following matrix:

$$\mathbf{G} = \begin{bmatrix} 7 & 0 & 4 \\ 7 & 0 & 4 \\ 0 & 0 & 11 \end{bmatrix}$$

6.2–8. Solve Problem 6.2–1 for the matrix **H**:

$$\mathbf{H} = \begin{bmatrix} -61 & 48 & 36 \\ -136 & 105 & 76 \\ 72 & -54 & -37 \end{bmatrix}$$

6.3–1. Determine the eigenvalues, a set of corresponding eigenvectors, and the number of linearly independent eigenvectors for the following matrix having repeated eigenvalues:

$$\mathbf{A} = \begin{bmatrix} -2 & 0 & -14 \\ -7 & 5 & -14 \\ 0 & 0 & 5 \end{bmatrix}$$

6.3–2. Solve Problem 6.3–1 for the matrix **B**:

$$\mathbf{B} = \begin{bmatrix} 77 & -60 & 15 \\ 150 & -118 & 30 \\ 200 & -160 & 42 \end{bmatrix}$$

6.3–3. Solve Problem 6.3–1 for the upper triangular matrix **C**:

$$\mathbf{C} = \begin{bmatrix} 3 & -5 & 1 \\ 0 & 3 & 0 \\ 0 & 0 & 3 \end{bmatrix}$$

6.3–4. Solve Problem 6.3–1 for the following matrix:

$$\mathbf{D} = \begin{bmatrix} 1 & 1 & 0 \\ 0 & 1 & 1 \\ 0 & 0 & 1 \end{bmatrix}$$

6.3–5. Solve Problem 6.3–1 for the following matrix:

$$\mathbf{E} = \begin{bmatrix} 0 & 1 & 0 \\ 0 & 0 & 1 \\ 1 & -3 & 3 \end{bmatrix}$$

6.4–1. Find the eigenvalues and a set of corresponding eigenvectors for the following matrix **A** and its transpose $\mathbf{A}^T$:

$$\mathbf{A} = \begin{bmatrix} -2 & -3 \\ 4 & 5 \end{bmatrix}$$

6.4–2. Solve the preceding problem for the following matrix **B** and its transpose $\mathbf{B}^T$:

$$\mathbf{B} = \begin{bmatrix} 3 & 0 & -8 \\ 2 & 1 & -4 \\ 0 & 0 & -1 \end{bmatrix}$$

6.4–3. Determine the eigenvalues and a set of corresponding eigenvectors for the following matrix **C** and its inverse $\mathbf{C}^{-1}$:

$$\mathbf{C} = \begin{bmatrix} 2 & -2 \\ 0 & 3 \end{bmatrix}$$

6.4–4. Solve the preceding problem for the matrix **D** shown below and its inverse $\mathbf{D}^{-1}$:

$$\mathbf{D} = \begin{bmatrix} 5 & -18 \\ 0 & -4 \end{bmatrix}$$

6.4–5. Solve Problem 6.4–3 for the matrix **E** and its inverse $\mathbf{E}^{-1}$:

$$\mathbf{E} = \begin{bmatrix} 1 & -1 & 0 \\ 2 & 4 & 0 \\ -2 & -1 & 1 \end{bmatrix}$$

6.4–6. Using the matrix **E** of the preceding problem, find the eigenvalues and a set of corresponding eigenvectors for the matrix 3**E**.

6.4–7. Determine the eigenvalues and a set of corresponding eigenvectors for the following orthogonal matrix:

$$\mathbf{F} = \begin{bmatrix} 0.6 & 0.8 \\ 0.8 & -0.6 \end{bmatrix}$$

6.4–8. Solve the preceding problem for the orthogonal matrix **G**:

$$\mathbf{G} = \begin{bmatrix} \frac{1}{2} & \frac{\sqrt{3}}{2} \\ \frac{\sqrt{3}}{2} & -\frac{1}{2} \end{bmatrix}$$

6.4–9. Repeat Problem 6.4–7 for the orthogonal matrix **H**:

$$\mathbf{H} = \begin{bmatrix} \frac{5}{13} & \frac{12}{13} & 0 \\ \frac{12}{13} & -\frac{5}{13} & 0 \\ 0 & 0 & 1 \end{bmatrix}$$

6.4–10. Solve Problem 6.4–7 for the orthogonal matrix **J**:

$$\mathbf{J} = \begin{bmatrix} \frac{1}{2} & \frac{1}{2} & \frac{\sqrt{2}}{2} \\ \frac{1}{2} & \frac{1}{2} & -\frac{\sqrt{2}}{2} \\ \frac{\sqrt{2}}{2} & -\frac{\sqrt{2}}{2} & 0 \end{bmatrix}$$

6.4–11. Determine the eigenvalues and a set of corresponding eigenvectors for the upper triangular matrix **K**:

$$\mathbf{K} = \begin{bmatrix} 2 & -4 & 1 \\ 0 & 3 & 2 \\ 0 & 0 & -1 \end{bmatrix}$$

6.4–12. Solve the preceding problem for the lower triangular matrix **L**:

$$\mathbf{L} = \begin{bmatrix} 6 & 0 & 0 \\ 3 & 0 & 0 \\ 2 & 1 & -10 \end{bmatrix}$$

6.4–13. Solve Problem 6.4–11 for the fourth-order upper triangular matrix shown below:

$$\mathbf{M} = \begin{bmatrix} 6 & 12 & 0 & 2 \\ 0 & 3 & 1 & 2 \\ 0 & 0 & -2 & 0 \\ 0 & 0 & 0 & 1 \end{bmatrix}$$

6.4–14. Determine the eigenvalues and a set of corresponding eigenvectors for the following diagonal matrix in which d_1, d_2, and d_3 are distinct:

$$\mathbf{N} = \begin{bmatrix} d_1 & 0 & 0 \\ 0 & d_2 & 0 \\ 0 & 0 & d_3 \end{bmatrix}$$

6.4–15. Using the following matrices **A** and **B**, show that the matrix products **AB** and **BA** have the same eigenvalues:

$$\mathbf{A} = \begin{bmatrix} 8 & -1 \\ -12 & 1 \end{bmatrix} \qquad \mathbf{B} = \begin{bmatrix} -1 & -1 \\ 3 & 2 \end{bmatrix}$$

6.4–16. Verify Equations (6–29) and (6–30) for the matrices **A** and **B** of Problems 6.4–1 and 6.4–2.

6.5–1. Diagonalize the 2 X 2 matrix **A** used in Section 6.1 as a numerical example.

6.5–2. Diagonalize the 3 X 3 matrix **B** used in Section 6.1 as a numerical example.

6.5–3. Diagonalize the 2 X 2 matrix **C** given in Problem 6.2–3.

6.5–4. Diagonalize the 2 X 2 matrix **D** given in Problem 6.4–4.

6.5–5. Diagonalize the triangular matrix **E** shown below:

$$\mathbf{E} = \begin{bmatrix} -2 & 0 & 0 \\ 5 & 2 & 0 \\ 1 & 4 & -3 \end{bmatrix}$$

6.5–6. Diagonalize the matrix **F** given in Problem 6.2–6.

6.5–7. Diagonalize the orthogonal matrix **G** given in Problem 6.4–8.

6.5–8. Diagonalize the matrix **H** shown below:

$$\mathbf{H} = \begin{bmatrix} 0 & 0 & 9 \\ 0 & 7 & 0 \\ 1 & 0 & 0 \end{bmatrix}$$

6.5–9. Diagonalize the matrix **J** shown below (see Problem 6.3–2):

$$\mathbf{J} = \begin{bmatrix} 77 & -60 & 15 \\ 150 & -118 & 30 \\ 200 & -160 & 42 \end{bmatrix}$$

6.5–10. Show that the following matrices cannot be diagonalized:

$$\mathbf{K}_1 = \begin{bmatrix} 1 & 4 \\ 0 & 1 \end{bmatrix} \qquad \mathbf{K}_2 = \begin{bmatrix} -2 & -1 \\ 1 & 0 \end{bmatrix}$$

6.5–11. Show that the following matrix cannot be diagonalized:

$$\mathbf{L} = \begin{bmatrix} 3 & 0 & 0 \\ -5 & 3 & 0 \\ 2 & 1 & 3 \end{bmatrix}$$

6.7–1. Determine the eigenvalues and a set of corresponding eigenvectors for the symmetric matrix **A**:

$$\mathbf{A} = \begin{bmatrix} 2 & 0 & 0 \\ 0 & 10 & 6 \\ 0 & 6 & 5 \end{bmatrix}$$

6.7–2. Repeat the preceding problem for the following matrix **B**:

$$\mathbf{B} = \begin{bmatrix} -23 & 0 & 28 \\ 0 & 17 & 0 \\ 28 & 0 & 10 \end{bmatrix}$$

6.7–3. Solve Problem 6.7–1 for the matrix **C**:

$$\mathbf{C} = \begin{bmatrix} 0 & 0 & 81 \\ 0 & 64 & 0 \\ 81 & 0 & 0 \end{bmatrix}$$

6.7–4. Solve Problem 6.7–1 for the matrix **D**:

$$\mathbf{D} = \begin{bmatrix} -48 & 36 & 0 \\ 36 & -27 & 0 \\ 0 & 0 & 50 \end{bmatrix}$$

6.7–5. Solve Problem 6.7–1 for the matrix **E**:

$$\mathbf{E} = \begin{bmatrix} 8 & -2 & \sqrt{2} \\ -2 & 8 & \sqrt{2} \\ \sqrt{2} & \sqrt{2} & 6 \end{bmatrix}$$

6.7–6. Solve Problem 6.7–1 for the matrix **F**:

$$\mathbf{F} = \begin{bmatrix} 3 & 5 & 0 \\ 5 & 3 & 0 \\ 0 & 0 & 8 \end{bmatrix}$$

6.7–7. Solve Problem 6.7–1 for the matrix **G**:

$$\mathbf{G} = \begin{bmatrix} -30 & 60 & 0 \\ 60 & 5 & 0 \\ 0 & 0 & 50 \end{bmatrix}$$

6.7–8. Solve Problem 6.7–1 for the matrix **H**:

$$\mathbf{H} = \begin{bmatrix} 10 & -10 & 0 \\ -10 & 10 & 0 \\ 0 & 0 & 0 \end{bmatrix}$$

6.7–9. Solve Problem 6.7–1 for the matrix **J**:

$$\mathbf{J} = \begin{bmatrix} 3 & 3 & 3\sqrt{2} \\ 3 & 3 & 3\sqrt{2} \\ 3\sqrt{2} & 3\sqrt{2} & 6 \end{bmatrix}$$

6.7–10. Solve Problem 6.7–1 for the matrix **K**:

$$\mathbf{K} = \begin{bmatrix} 6 & 2 & 2\sqrt{2} \\ 2 & 6 & 2\sqrt{2} \\ 2\sqrt{2} & 2\sqrt{2} & 8 \end{bmatrix}$$

6.7–11 to 6.7–20. Diagonalize the matrices **A, B, C, D, E, F, G, H, J,** and **K** given in Problems 6.7–1 to 6.7–10, respectively.

6.8–1. Assuming that **B** is any square matrix having a square root $\sqrt{\mathbf{B}}$, show that $\mathbf{A}\sqrt{\mathbf{B}}\,\mathbf{A}^{-1}$ is a square root of $\mathbf{ABA}^{-1}$, where **A** is a nonsingular square matrix of the same order as **B**.

6.8–2. Determine the four square roots of the following matrix:

$$\mathbf{A} = \begin{bmatrix} 5 & 4 \\ 4 & 5 \end{bmatrix}$$

6.8–3. Determine the four square roots of the following matrix:

$$\mathbf{A} = \begin{bmatrix} 5 & 8 \\ 8 & 13 \end{bmatrix}$$

(Hint: To simplify the arithmetic, note that $\sqrt{9 + 4\sqrt{5}} = 2 + \sqrt{5}$ and $\sqrt{9 - 4\sqrt{5}} = 2 - \sqrt{5}$.)

6.8–4. Determine the eight square roots of the following matrix by utilizing Equation (6–55) and the eigenvalues and eigenvectors calculated in Example 1 of Section 6.7:

$$\mathbf{A} = \begin{bmatrix} 2 & 2 & 0 \\ 2 & 5 & 0 \\ 0 & 0 & 3 \end{bmatrix}$$

6.8–5. Determine a square root of the following unsymmetric matrix:

$$\mathbf{A} = \begin{bmatrix} 111 & -59 & 6 \\ -60 & 86 & -30 \\ -20 & -107 & 89 \end{bmatrix}$$

Note: This matrix has eigenvalues 176, 99, and 11 and a modal matrix

$$\mathbf{M} = \begin{bmatrix} 1 & -1 & 1 \\ -1 & 0 & 2 \\ 1 & 2 & 3 \end{bmatrix}$$

6.8–6. Determine a square root of the matrix **A** given in Problem 6.7–1.

6.8–7. Determine the fifth power of the matrix **A** given by Equation (6–6) in Section 6.1.

6.8–8. Determine the cube of the matrix **A** given by Equation (6–44) in Section 6.7.

6.9–1. Determine the eigenvalues and a set of corresponding eigenvectors for the symmetric matrix **A** given below using the iteration method:

$$\mathbf{A} = \begin{bmatrix} 4 & 2 & 1 \\ 2 & 7 & 3 \\ 1 & 3 & 5 \end{bmatrix}$$

The numerically largest eigenvalue and associated eigenvector are approximately as follows:

$$\lambda_1 \approx 10 \qquad \mathbf{X}_1 \approx \beta_1 \{1 \quad 2 \quad 1.5\}$$

6.9–2. Determine the eigenvalues and a set of corresponding eigenvectors for the matrix **B** using iteration:

$$\mathbf{B} = \begin{bmatrix} 12 & 7 & 5 \\ 7 & 4 & 9 \\ 5 & 9 & 2 \end{bmatrix}$$

The numerically largest eigenvalue and associated eigenvector are, approximately,

$$\lambda_1 \approx 20 \qquad \mathbf{X}_1 \approx \beta_1 \{1 \quad 1 \quad 1\}$$

6.9–3. Determine the eigenvalues and a set of corresponding eigenvectors for the following matrix using iteration:

$$\mathbf{C} = \begin{bmatrix} 476 & 293 & 270 \\ 293 & 814 & 157 \\ 270 & 157 & 684 \end{bmatrix}$$

Use $\mathbf{X}_{(1)} = \{1 \quad 1 \quad 1\}$ as a trial eigenvector.

6.9–4. Find the eigenvalues and a set of corresponding eigenvectors for the matrix **D** using iteration:

$$\mathbf{D} = \begin{bmatrix} 17.8 & 29.3 & 11.5 \\ 29.3 & 20.6 & 18.4 \\ 11.5 & 18.4 & 12.4 \end{bmatrix}$$

Assume a trial vector $\mathbf{X}_{(1)} = \{1 \quad 1 \quad 1\}$.

7

...Coordinate Transformations...

Many kinds of technical analyses require the use of two or more coordinate systems. For instance, in engineering mechanics problems it is often necessary to make use of a coordinate system that is rotated with respect to a fixed reference system. Some techniques for handling transformations between coordinate systems are described in this chapter.

7.1. Rotation of Axes for Vectors. The first type of problem to be considered is the rotation of coordinate axes when only three-dimensional vector quantities are involved. For this purpose, a vector **V** directed from the origin O to a fixed point P in three-dimensional space is shown in Figure 7-1a. For convenience in notation, the orthogonal reference axes are denoted as x_1, x_2, and x_3 instead of x, y, and z. The components of the vector **V** in the directions of these coordinate axes are X_1, X_2, and X_3 (also shown in the figure). When the vector starts at the origin, as in Figure 7-1a, the components of the vector are also the coordinates of the point where the vector terminates. Hence, it is permissible to consider the components of the vector **V** as the coordinates of point P.

A second set of axes y_1, y_2, and y_3 is shown in Figure 7-1b. These axes have the same origin as the first set of axes, but they are rotated with respect to the first set. The fixed point P and the vector **V** appear again in this figure, and the components of **V** with respect to the second coordinate system are Y_1, Y_2, and Y_3. Thus, the vector **V** (which is the same vector in both parts of Figure 7-1) may be represented either by its components in the first system or by its components in the second system. When it is represented in the first system, the vector will be denoted by the symbol **X**; and when it is represented in

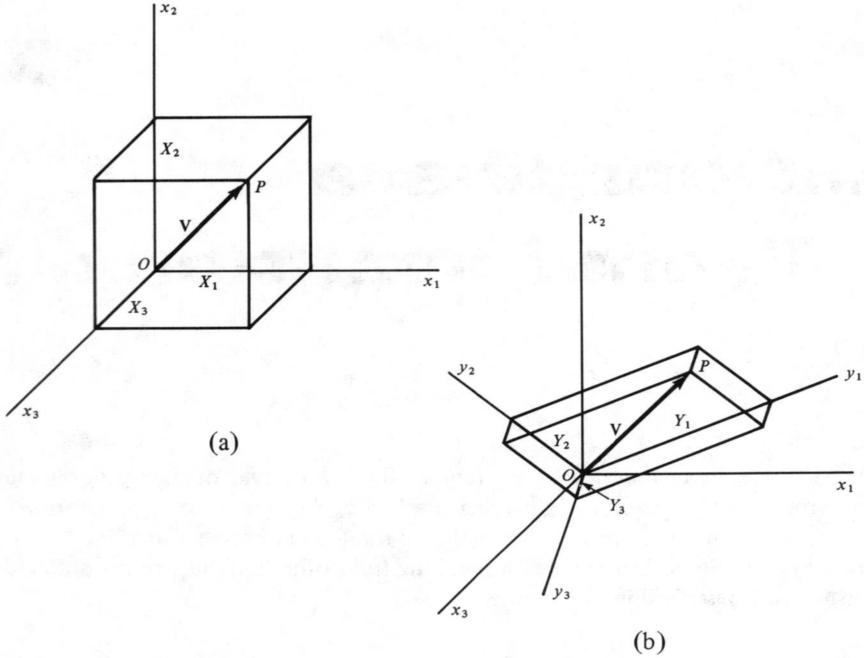

Figure 7-1. Components of vector **V** in: (a) first system of axes, (b) second system of axes

the second system, the symbol **Y** will be used. Thus, the following column vectors

$$\mathbf{X} = \begin{bmatrix} X_1 \\ X_2 \\ X_3 \end{bmatrix} \qquad \mathbf{Y} = \begin{bmatrix} Y_1 \\ Y_2 \\ Y_3 \end{bmatrix} \tag{7-1}$$

are alternative representations of the same vector **V**.

The most convenient way to describe the relative positions of the two coordinate systems shown in Figure 7-1b is by means of *direction cosines*. A direction cosine is defined as the cosine of the angle between the positive directions of two coordinate axes. The direction cosines of the y_1 axis with respect to the x_1, x_2, and x_3 axes will be denoted L_{11}, L_{12}, and L_{13}, respectively. Similarly, the direction cosines of the axes y_2 and y_3 with respect to x_1, x_2, and x_3 will be denoted L_{21}, L_{22}, L_{23} and L_{31}, L_{32}, L_{33}, respectively. The subscript notation for the direction cosines is as follows: the first subscript refers to the second system of axes, and the second subscript refers to the first system. By this convention, L_{ij} is the cosine of the angle between the ith axis of the second system and the jth axis of the first system.

The direction cosines serve to relate the components of the vector **V** in the second (or y) system to those in the first (or x) system, as expressed in the following equations:

$$\begin{aligned} Y_1 &= L_{11}X_1 + L_{12}X_2 + L_{13}X_3 \\ Y_2 &= L_{21}X_1 + L_{22}X_2 + L_{23}X_3 \\ Y_3 &= L_{31}X_1 + L_{32}X_2 + L_{33}X_3 \end{aligned} \tag{7-2}$$

In matrix form, these equations are

$$\mathbf{Y} = \mathbf{RX} \tag{7-3}$$

in which **X** and **Y** are the vectors given by Equations (7-1), and **R** is the matrix of direction cosines, called the *rotation matrix*:

$$\mathbf{R} = \begin{bmatrix} L_{11} & L_{12} & L_{13} \\ L_{21} & L_{22} & L_{23} \\ L_{31} & L_{32} & L_{33} \end{bmatrix} \tag{7-4}$$

Note that the ith row of the rotation matrix consists of the direction cosines of the ith axis of the second system with respect to the three axes of the first system.

An interesting property of the rotation matrix can be observed by recalling from analytic geometry some relationships among the direction cosines. First, the sum of the squares of the direction cosines for any axis is equal to unity, as shown:

$$\begin{aligned} L_{11}^2 + L_{12}^2 + L_{13}^2 &= 1 \\ L_{21}^2 + L_{22}^2 + L_{23}^2 &= 1 \\ L_{31}^2 + L_{32}^2 + L_{33}^2 &= 1 \end{aligned} \tag{7-5}$$

Second, when two axes are orthogonal to one another, the sum of the products of corresponding direction cosines is zero. This relation gives the following equations for the y_1y_2, y_1y_3, and y_2y_3 axes, respectively:

$$\begin{aligned} L_{11}L_{21} + L_{12}L_{22} + L_{13}L_{23} &= 0 \\ L_{11}L_{31} + L_{12}L_{32} + L_{13}L_{33} &= 0 \\ L_{21}L_{31} + L_{22}L_{32} + L_{23}L_{33} &= 0 \end{aligned} \tag{7-6}$$

Equations (7-5) and (7-6) can be expressed more concisely by considering the rows of the rotation matrix **R** to be three row vectors $\mathbf{L}_1$, $\mathbf{L}_2$, and $\mathbf{L}_3$, respectively:

$$\mathbf{L}_1 = [L_{11} \quad L_{12} \quad L_{13}] \qquad \mathbf{L}_2 = [L_{21} \quad L_{22} \quad L_{23}] \qquad \mathbf{L}_3 = [L_{31} \quad L_{32} \quad L_{33}] \tag{7-7}$$

Equations (7-5) denote the fact that the vectors $\mathbf{L}_1$, $\mathbf{L}_2$, and $\mathbf{L}_3$ each have unit length, and Equations (7-6) convey the fact that the scalar product of one vector with another is zero; that is, the vectors are orthogonal. Thus, the three vectors $\mathbf{L}_1$, $\mathbf{L}_2$, and $\mathbf{L}_3$, which are in the directions of the y_1, y_2, and y_3 axes, respectively, constitute a set of orthonormal vectors. These vectors satisfy the equation

$$\mathbf{L}_i\mathbf{L}_j^{\mathsf{T}} = \delta_{ij} \tag{7-8}$$

in which δ_{ij} is the Kronecker delta (see Equation 4-24). Because the rows of the rotation matrix are a set of orthonormal vectors, it follows that $\mathbf{R}$ is an orthogonal matrix, as explained previously in Section 4.6.

Most of the preceding equations were based upon the assumption that the vector $\mathbf{Y}$ was to be expressed in terms of the vector $\mathbf{X}$ (see Equation 7-3). Consider now the reverse problem of relating the components in the first system to those in the second; that is, expressing the vector $\mathbf{X}$ in terms of the vector $\mathbf{Y}$. For this purpose, the direction cosines defined above may serve also as direction cosines of the first system with respect to the second system, inasmuch as they are simply the cosines of the angles between pairs of axes. Hence, the direction cosines of the x_1 axis with respect to the y_1, y_2, and y_3 axes are L_{11}, L_{21}, and L_{31}, respectively. Also the direction cosines for the x_2 and x_3 axes are L_{12}, L_{22}, L_{32}, and L_{13}, L_{23}, L_{33}, respectively. Thus, the components of the vector $\mathbf{V}$ in the first system are related to those in the second system by the equations

$$\begin{aligned} X_1 &= L_{11}Y_1 + L_{21}Y_2 + L_{31}Y_3 \\ X_2 &= L_{12}Y_1 + L_{22}Y_2 + L_{32}Y_3 \\ X_3 &= L_{13}Y_1 + L_{23}Y_2 + L_{33}Y_3 \end{aligned} \tag{7-9}$$

or, in matrix form,

$$\mathbf{X} = \mathbf{R}^{\mathsf{T}}\mathbf{Y} \tag{7-10a}$$

in which $\mathbf{R}^{\mathsf{T}}$ is the transpose of the rotation matrix $\mathbf{R}$. Since the rotation matrix is an orthogonal matrix, its inverse is the same as its transpose, and therefore Equation (7-10a) can also be written in the form

$$\mathbf{X} = \mathbf{R}^{-1}\mathbf{Y} \tag{7-10b}$$

which is in agreement with Equation (7-3). It was pointed out earlier in Section 4.6 that not only the rows but also the columns of an orthogonal matrix constitute an orthonormal set of vectors. Hence, six additional equations analogous to Equations (7-5) and (7-6) can be written for the direction cosines appearing in the rows of $\mathbf{R}^{\mathsf{T}}$, which are the columns of $\mathbf{R}$.

Successive rotations of axes for vectors can be accomplished by successive multiplications of rotation matrices. As an illustration, suppose that a rotation matrix $\mathbf{R}_1$ transforms a vector $\mathbf{X}$ from the first system of axes to a vector $\mathbf{Y}$ in the second system, and that a rotation matrix $\mathbf{R}_2$ transforms $\mathbf{Y}$ in the second system to $\mathbf{Z}$ in a third system. Then the vector $\mathbf{Y}$ is related to the vector $\mathbf{X}$ by the equation

$$\mathbf{Y} = \mathbf{R}_1\mathbf{X} \tag{a}$$

and the vector **Z** is related to the vector **Y** as follows:

$$\mathbf{Z} = \mathbf{R}_2 \mathbf{Y} \tag{b}$$

Substitution of **Y** from Equation (a) into Equation (b) produces the vector **Z** in terms of **X**:

$$\mathbf{Z} = \mathbf{R}_2 \mathbf{R}_1 \mathbf{X} = \mathbf{R}_3 \mathbf{X} \tag{c}$$

The rotation matrix $\mathbf{R}_3$, which is obtained as the product of the first two rotation matrices, is itself an orthogonal matrix (see Equation 4-26). The concept of successive multiplication of rotation matrices holds for any number of successive rotations of axes, and the product matrix will be the rotation matrix that directly transforms the vector from the first system of axes to the last one.

One final property of the rotation matrix that distinguishes it from orthogonal matrices in general is that its determinant is always equal to plus one. This property has occasional usefulness as a check when performing numerical calculations.

Two-Dimensional Rotation of Axes. A special case of rotation of axes occurs when the vector **V** lies in one of the three coordinate planes of the $x_1x_2x_3$ system and the rotation occurs about the axis which is perpendicular to that plane. Assume, for instance, that the vector **V** lies in the x_1x_2 plane (see Figure 7-2). The second system of axes is obtained by rotating through an angle θ about the x_3 axis, which is perpendicular to the x_1x_2 plane as shown in the sketch. Thus, the axes y_1 and y_2 are in the x_1x_2 plane, and the y_3 axis coincides with the x_3 axis. A rotation of axes of this kind frequently is called a two-dimensional rotation of axes, or rotation of axes in two dimensions.

For the rotation of axes shown in Figure 7-2, the direction cosines are as follows:

$$L_{11} = \cos\theta \quad L_{12} = \cos(90 - \theta) = \sin\theta \quad L_{13} = \cos 90^\circ = 0$$
$$L_{21} = \cos(90 + \theta) = -\sin\theta \quad L_{22} = \cos\theta \quad L_{23} = \cos 90^\circ = 0$$
$$L_{31} = \cos 90^\circ = 0 \quad L_{32} = \cos 90^\circ = 0 \quad L_{33} = \cos 0^\circ = 1$$

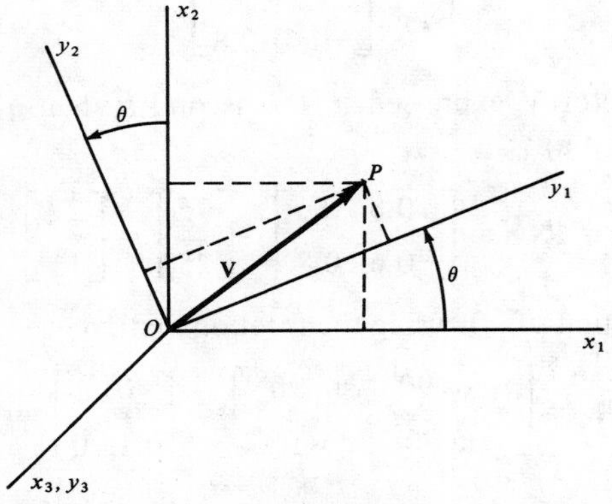

Figure 7-2. Two-dimensional rotation of axes

Hence, the rotation matrix **R** is

$$\mathbf{R} = \begin{bmatrix} \cos\theta & \sin\theta & 0 \\ -\sin\theta & \cos\theta & 0 \\ 0 & 0 & 1 \end{bmatrix} \tag{7-11}$$

If the third components of both **X** and **Y** are dropped from consideration because they are always zero in a two-dimensional problem, these vectors simplify to

$$\mathbf{X} = \begin{bmatrix} X_1 \\ X_2 \end{bmatrix} \qquad \mathbf{Y} = \begin{bmatrix} Y_1 \\ Y_2 \end{bmatrix} \tag{7-12}$$

and the rotation matrix is reduced to the portion which remains when the third row and the third column are deleted from **R** in Equation (7-11):

$$\mathbf{R} = \begin{bmatrix} \cos\theta & \sin\theta \\ -\sin\theta & \cos\theta \end{bmatrix} \tag{7-13}$$

Thus, for rotation of axes in two dimensions, Equations (7-12) and (7-13) can be used in place of their three-dimensional counterparts.

Example. As a numerical example, suppose that the vector **V** when expressed in the first coordinate system is

$$\mathbf{X} = \begin{bmatrix} X_1 \\ X_2 \end{bmatrix} = \begin{bmatrix} 15 \\ 25 \end{bmatrix}$$

and that the first rotation angle θ is such that

$$\tan\theta = \frac{3}{4}$$

It follows that $\theta = 36^\circ\ 52'$, $\sin\theta = 0.6$, and $\cos\theta = 0.8$; hence, the rotation matrix given by Equation (7-13) is

$$\mathbf{R}_1 = \begin{bmatrix} 0.8 & 0.6 \\ -0.6 & 0.8 \end{bmatrix}$$

Therefore, the vector **V** expressed in the second system of coordinates becomes (see Equation a)

$$\mathbf{Y} = \mathbf{R}_1\mathbf{X} = \begin{bmatrix} 0.8 & 0.6 \\ -0.6 & 0.8 \end{bmatrix} \begin{bmatrix} 15 \\ 25 \end{bmatrix} = \begin{bmatrix} 27 \\ 11 \end{bmatrix}$$

A second rotation of 90° using the rotation matrix

$$\mathbf{R}_2 = \begin{bmatrix} \cos 90^\circ & \sin 90^\circ \\ -\sin 90^\circ & \cos 90^\circ \end{bmatrix} = \begin{bmatrix} 0 & 1 \\ -1 & 0 \end{bmatrix}$$

gives the vector **V** in a third coordinate system, as indicated by Equations (b) and (c). Therefore,

$$\mathbf{Z} = \mathbf{R}_2\mathbf{R}_1\mathbf{X} = \mathbf{R}_3\mathbf{X} = \begin{bmatrix} -0.6 & 0.8 \\ -0.8 & -0.6 \end{bmatrix} \begin{bmatrix} 15 \\ 25 \end{bmatrix} = \begin{bmatrix} 11 \\ -27 \end{bmatrix}$$

Finally, a rotation back to the first set of axes produces the original vector (see Equation 7-10a):

$$\mathbf{X} = \mathbf{R}_3^T\mathbf{Z} = \begin{bmatrix} -0.6 & -0.8 \\ 0.8 & -0.6 \end{bmatrix} \begin{bmatrix} 11 \\ -27 \end{bmatrix} = \begin{bmatrix} 15 \\ 25 \end{bmatrix}$$

7.2. Rotation of Axes for Matrices. In the preceding section, transformation of a vector from one orthogonal three-dimensional coordinate system to another was described. It is sometimes necessary to transform a square matrix in an analogous manner. Assume, for instance, that an equation relating two vectors $\mathbf{X}_1$ and $\mathbf{X}_2$ in the same coordinate system has the form

$$\mathbf{A}\mathbf{X}_1 = \mathbf{X}_2 \tag{7-14}$$

in which **A** is a 3×3 matrix and $\mathbf{X}_1$ and $\mathbf{X}_2$ are three-dimensional vectors:

$$\mathbf{A} = \begin{bmatrix} A_{11} & A_{12} & A_{13} \\ A_{21} & A_{22} & A_{23} \\ A_{31} & A_{32} & A_{33} \end{bmatrix} \qquad \mathbf{X}_1 = \begin{bmatrix} X_{11} \\ X_{12} \\ X_{13} \end{bmatrix} \qquad \mathbf{X}_2 = \begin{bmatrix} X_{21} \\ X_{22} \\ X_{23} \end{bmatrix} \tag{7-15}$$

Note that two subscripts are used for each component of $\mathbf{X}_1$ and $\mathbf{X}_2$; the first subscript may be considered to be a part of the identifier itself and is the number of the vector, whereas the second subscript denotes the coordinate axis along which the component is measured. If it is desired to express Equation (7-14) in a second coordinate system (for example, the y axes shown in Figure 7-1b), then the transformation of the vectors $\mathbf{X}_1$ and $\mathbf{X}_2$ to the new system results in a corresponding transformation on the matrix **A**.

Equation (7-3) can be used to transform the vectors $\mathbf{X}_1$ and $\mathbf{X}_2$ to the second system of axes; therefore, in the rotated system they become

$$\mathbf{Y}_1 = \begin{bmatrix} Y_{11} \\ Y_{12} \\ Y_{13} \end{bmatrix} = \mathbf{R}\mathbf{X}_1 \qquad \mathbf{Y}_2 = \begin{bmatrix} Y_{21} \\ Y_{22} \\ Y_{23} \end{bmatrix} = \mathbf{R}\mathbf{X}_2 \tag{7-16}$$

in which **R** is the rotation matrix. The corresponding transformation for the matrix **A** may be obtained by premultiplying both sides of Equation (7-14) by **R** and also inserting the product $\mathbf{R}^{-1}\mathbf{R}$ (equal to **I**) between the matrix **A** and the vector $\mathbf{X}_1$ on the left-hand side; thus,

$$\mathbf{R}\mathbf{A}\mathbf{R}^{-1}\mathbf{R}\mathbf{X}_1 = \mathbf{R}\mathbf{X}_2$$

Introducing the notation

$$\mathbf{B} = \mathbf{RAR}^{-1} = \mathbf{RAR}^{\mathsf{T}} \tag{7-17}$$

the above equation may be written

$$\mathbf{BY}_1 = \mathbf{Y}_2 \tag{7-18}$$

Equation (7–17) shows the transformation of the matrix **A** into a new matrix **B** because of the rotation of axes, and Equation (7–18) represents the resulting transformed equations.

The reverse transformation from the second coordinate system to the first is accomplished by the equations

$$\mathbf{X}_1 = \mathbf{R}^{-1}\mathbf{Y}_1 = \mathbf{R}^{\mathsf{T}}\mathbf{Y}_1 \qquad \mathbf{X}_2 = \mathbf{R}^{-1}\mathbf{Y}_2 = \mathbf{R}^{\mathsf{T}}\mathbf{Y}_2 \tag{7-19}$$

$$\mathbf{A} = \mathbf{R}^{-1}\mathbf{BR} = \mathbf{R}^{\mathsf{T}}\mathbf{BR} \tag{7-20}$$

as can be obtained readily from Equations (7–16) and (7–17).

An important aspect to be kept in mind when performing rotation of axes is that the vectors themselves are considered to remain unchanged. The vectors $\mathbf{V}_1$ and $\mathbf{V}_2$ remain fixed in three-dimensional space, and they represent the same intrinsic quantities irrespective of the coordinate system. Thus, the vectors $\mathbf{X}_1$ and $\mathbf{X}_2$ in Equation (7–14) consist of the components of $\mathbf{V}_1$ and $\mathbf{V}_2$ in the x coordinate system, while the vectors $\mathbf{Y}_1$ and $\mathbf{Y}_2$ in Equation (7–18) consist of the components of $\mathbf{V}_1$ and $\mathbf{V}_2$ in the y coordinates. For the first system of axes, the relationships between the elements of $\mathbf{X}_1$ and $\mathbf{X}_2$ are expressed by the matrix **A**, while in the second system the relationships between the elements of $\mathbf{Y}_1$ and $\mathbf{Y}_2$ are expressed by the matrix **B**. The transformation of **A** into **B** denoted by Equation (7–17) may be interpreted as an operation that is necessary in order to retain the intrinsic relationships between the vectors $\mathbf{V}_1$ and $\mathbf{V}_2$. Furthermore, the transformation of the matrix **A** to the matrix **B** is accomplished by an orthogonal similarity transformation; hence, **A** and **B** have the same eigenvalues. Also, the eigenvectors of **B** are related to those of **A** by the vector transformation given in Equation (7-3), as can be observed by referring to Equations (6-41), (6-42), and (7–17).

7.3. Principal Axes. In the preceding sections, it was assumed that the $y_1y_2y_3$ axes were rotated to any arbitrary position with respect to the $x_1x_2x_3$ axes. Consider now a special set of rotated axes, called *principal axes*, for which the transformed matrix **B** in Equation (7–17) is a diagonal matrix. If the matrix **A** in that equation is symmetric, it can always be diagonalized by an orthogonal similarity transformation in which the transformation matrix **Q** consists of the orthonormal set of eigenvectors of **A** (see Section 6.7). Thus, an important relationship exists between the rotation matrix **R** and the transformation matrix **Q** when the rotated axes are principal axes.

Comparison of Equations (6–52) and (7–17) shows that the transpose of **Q** is the rotation matrix for the principal axes; thus,

$$\mathbf{R}_P = \mathbf{Q}^{\mathsf{T}} \tag{7-21}$$

in which the symbol $\mathbf{R}_P$ is used for the rotation matrix defining the principal axes. Since the rows of a rotation matrix are the direction cosines for the axes, Equation (7–21) shows that the direction cosines for the principal axes are the normalized eigenvectors of **A**.

If the principal axes are denoted p_1, p_2, and p_3, corresponding to y_1, y_2, and y_3, respectively, then the transformation equations from the $x_1x_2x_3$ axes to the principal axes can be summarized as shown (compare with Equations 7–16, 7–17, and 7–18):

$$\mathbf{P}_1 = \begin{bmatrix} P_{11} \\ P_{12} \\ P_{13} \end{bmatrix} = \mathbf{Q}^{\mathsf{T}}\mathbf{X}_1 \qquad \mathbf{P}_2 = \begin{bmatrix} P_{21} \\ P_{22} \\ P_{23} \end{bmatrix} = \mathbf{Q}^{\mathsf{T}}\mathbf{X}_2 \tag{7-22}$$

$$\mathbf{Q}^{\mathsf{T}}\mathbf{A}\mathbf{Q} = \mathbf{S} \tag{7-23}$$

$$\mathbf{S}\mathbf{P}_1 = \mathbf{P}_2 \tag{7-24}$$

in which $\mathbf{P}_1$ and $\mathbf{P}_2$ represent the vectors $\mathbf{X}_1$ and $\mathbf{X}_2$ (or $\mathbf{V}_1$ and $\mathbf{V}_2$) with respect to the principal axes, and **S** is the spectral matrix for **A**. The last equation shows that the ratios of corresponding elements of the vectors $\mathbf{P}_2$ and $\mathbf{P}_1$ are the eigenvalues of **A**:

$$\frac{P_{21}}{P_{11}} = \lambda_1 \qquad \frac{P_{22}}{P_{12}} = \lambda_2 \qquad \frac{P_{23}}{P_{13}} = \lambda_3 \tag{7-25}$$

Now consider the eigenvectors of the matrix **A** and the manner in which they transform to the principal axes. The eigenvectors of **A** are the columns of **Q**; the transformation of these vectors to the principal axes is accomplished by premultiplying with the rotation matrix $\mathbf{R}_P$, which is equal to $\mathbf{Q}^{\mathsf{T}}$ (see Equation 7–21). Therefore, the eigenvectors with respect to the principal axes are the columns of the identity matrix. This same conclusion is obtained when the eigenvectors of the spectral matrix **S** are determined directly.

Example. The geometric significance of principal axes will be shown by means of a two-dimensional numerical example. Assume that a symmetric matrix **A** and two vectors $\mathbf{X}_1$ and $\mathbf{X}_2$, related as shown by Equation (7–14), are as follows:

$$\mathbf{A} = \begin{bmatrix} 1.95 & 0.15 \\ 0.15 & 1.55 \end{bmatrix} \qquad \mathbf{X}_1 = \begin{bmatrix} 1.00 \\ 1.60 \end{bmatrix} \qquad \mathbf{X}_2 = \begin{bmatrix} 2.19 \\ 2.63 \end{bmatrix}$$

The components of the vectors $\mathbf{X}_1$ and $\mathbf{X}_2$ may be considered to be the components of two vectors $\mathbf{V}_1$ and $\mathbf{V}_2$ that are expressed in terms of the x_1x_2 coordinate system. These vectors are plotted with respect to the x_1 and x_2 axes in Figure 7-3.

In order to find the principal axes, the eigenvalues and eigenvectors of $\mathbf{A}$ are needed. These quantities can be obtained by the methods described in Sections 6.1 and 6.2, and from them the spectral matrix $\mathbf{S}$ and a modal matrix $\mathbf{Q}$ of normalized eigenvectors are found:

$$\mathbf{S} = \begin{bmatrix} 2 & 0 \\ 0 & 1.5 \end{bmatrix} \qquad \mathbf{Q} = \frac{1}{\sqrt{10}} \begin{bmatrix} 3 & -1 \\ 1 & 3 \end{bmatrix}$$

The columns of $\mathbf{Q}$, which are the normalized eigenvectors of $\mathbf{A}$, are the direction cosines of the principal axes p_1 and p_2 with respect to the x_1x_2 axes (see Equation 7-21). The principal axes are also plotted in Figure 7-3.

Because each eigenvector of the matrix $\mathbf{A}$ can be multiplied by an arbitrary scalar, any one of the columns of $\mathbf{Q}$ can be multiplied by -1. When the direction cosines for an axis have their signs changed in this manner, the effect is to reverse the direction of the axis. When all such possible reversals in direction are taken into account, it is found that four different pairs of principal axes can be obtained, as shown in Figure 7-4. Part (a) of this figure shows the same principal axes as in Figure 7-3; these axes are determined by the matrix $\mathbf{Q}$ given above. Shown in the remaining parts of Figure 7-4 are the principal axes obtained by multiplying one or both of the columns of $\mathbf{Q}$ by -1. The columns of $\mathbf{Q}$ associated with each axis are labeled in the figures,

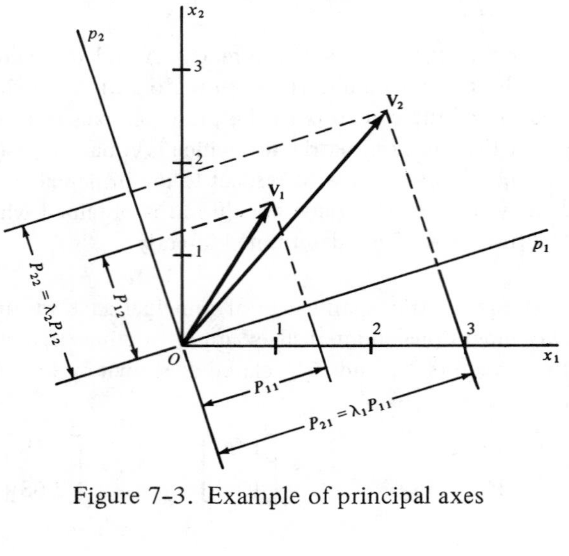

Figure 7-3. Example of principal axes

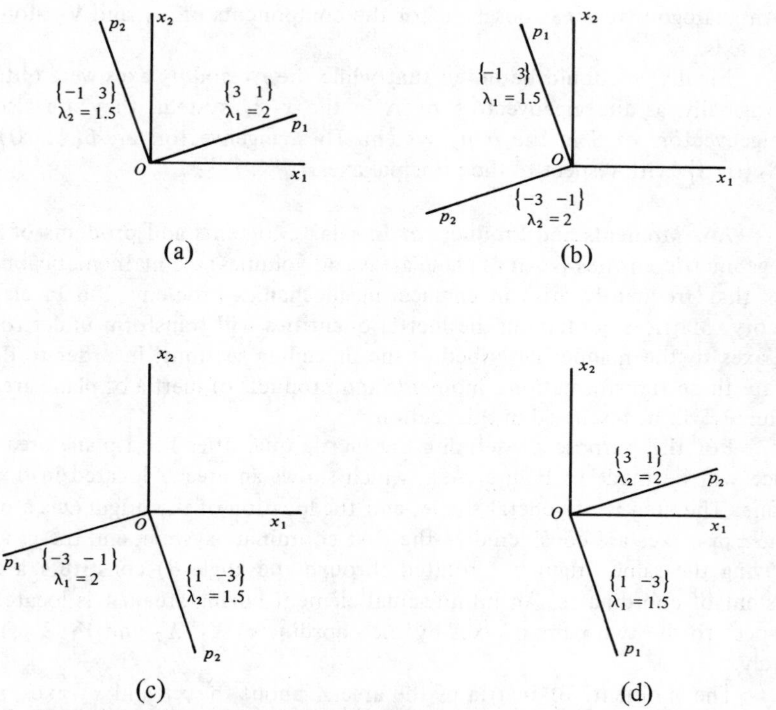

Figure 7-4. Example showing four positions of principal axes

except that the factor $1/\sqrt{10}$ has been omitted for convenience. The eigenvalues associated with the principal axes (that is, associated with the eigenvectors) are also shown in the figures. Each successive pair of principal axes is rotated 90° from the preceding pair, so that there is actually only one pair of mutually orthogonal principal axes. It is merely the directions of the axes that are affected by multiplying the normalized eigenvectors by -1.

The vectors $\mathbf{V}_1$ and $\mathbf{V}_2$ can be expressed in the principal coordinate system by using Equations (7-22); thus:

$$\mathbf{P}_1 = \mathbf{Q}^T\mathbf{X}_1 = \begin{bmatrix} P_{11} \\ P_{12} \end{bmatrix} = \begin{bmatrix} 1.45 \\ 1.20 \end{bmatrix} \qquad \mathbf{P}_2 = \mathbf{Q}^T\mathbf{X}_2 = \begin{bmatrix} P_{21} \\ P_{22} \end{bmatrix} = \begin{bmatrix} 2.90 \\ 1.80 \end{bmatrix}$$

The components of these vectors are the projections of $\mathbf{V}_1$ and $\mathbf{V}_2$ on the p_1 and p_2 axes, as indicated in Figure 7-3.

A significant feature of principal axes is that the ratios of corresponding components of the vectors are equal to the eigenvalues (see Equation 7-25). This fact is illustrated in Figure 7-3, where it can be seen that the component P_{21} of vector $\mathbf{V}_2$ is equal to λ_1 times the corresponding component of $\mathbf{V}_1$.

An analogous result is observed for the components of $\mathbf{V}_1$ and $\mathbf{V}_2$ along the p_2 axis.

Finally, it should be noted that while the p_1 and p_2 axes were obtained originally as the eigenvectors of $\mathbf{A}$ in the x_1x_2 system, they are also the eigenvectors of $\mathbf{S}$ in the p_1p_2 system. These eigenvectors are $\beta_1\{1 \quad 0\}$ and $\beta_2\{0 \quad 1\}$ with respect to the principal axes.

7.4. Moments and Products of Inertia. Moments and products of inertia of geometric entities (such as plane areas and volumes) are mathematical properties that frequently arise in engineering mechanics problems and in elasticity theory. Matrices containing the inertia quantities will transform under rotation of axes in the manner described in the preceding sections. In order to demonstrate these transformations, moments and products of inertia of plane areas and volumes will be discussed in this section.*

For the purpose of defining the inertia quantities for a plane area, reference will be made to Figure 7-5a, which shows an area A located in the x_1x_2 plane. The area is of general shape, and the location of the origin O is arbitrary. The x_1x_2 axes are considered as the first coordinate system, and the y_1y_2 axes (having the same origin but rotated through the angle θ) constitute a second system of coordinates. An infinitesimal element having area dA is located with respect to the two pairs of axes by the coordinates X_1, X_2 and Y_1, Y_2, respectively.

The moments of inertia of the area A about the x_1 and x_2 axes, respectively, are defined as

$$I_{X11} = \int X_2^2 dA \qquad I_{X22} = \int X_1^2 dA \tag{7-26}$$

in which it is assumed that the integrals are evaluated for the entire area. Also, the products of inertia with respect to the x_1x_2 axes are taken to be†

$$I_{X12} = -\int X_2 X_1 dA \qquad I_{X21} = -\int X_1 X_2 dA \tag{7-27}$$

*The terms "area" and "volume" are used in this section to mean "plane surface" and "solid," respectively. Even though the latter terms are correct, the former ones are so commonly used when referring to moments and products of inertia that they are used here also. Furthermore, even the term "inertia" is a misnomer when describing properties of areas and volumes, because the word inertia properly applies only to a body that has mass. Nevertheless, this terminology is universally used and will be followed here. The name arose historically in problems of dynamics, where the mass moments of inertia of solid bodies play an important role.

†The products of inertia are often defined without the minus signs shown in Equations (7-27). In that case, it is necessary to insert minus signs in many of the matrices and equations that follow.

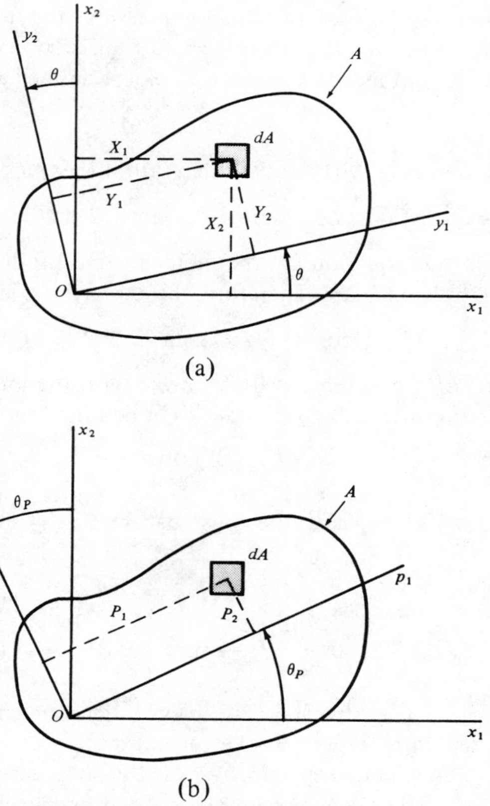

Figure 7-5. Plane area

It is evident from Equations (7-27) that the products of inertia I_{X12} and I_{X21} are equal:

$$I_{X12} = I_{X21} \tag{7-28}$$

Therefore, the symbol I_{X12} may be used to represent both of these quantities.

The relationships among the coordinates of the element dA in the two systems of axes are given by Equation (7-3), using the 2×2 rotation matrix of Equation (7-13):

$$\begin{bmatrix} Y_1 \\ Y_2 \end{bmatrix} = \mathbf{R} \begin{bmatrix} X_1 \\ X_2 \end{bmatrix} = \begin{bmatrix} \cos\theta & \sin\theta \\ -\sin\theta & \cos\theta \end{bmatrix} \begin{bmatrix} X_1 \\ X_2 \end{bmatrix}$$

or

$$Y_1 = X_1 \cos\theta + X_2 \sin\theta \tag{7-29a}$$

$$Y_2 = -X_1 \sin\theta + X_2 \cos\theta \tag{7-29b}$$

These equations can be used to obtain expressions for the moments and products of inertia of the area A with respect to the second system of axes. For instance, the moment of inertia about the y_1 axis is defined as

$$I_{Y11} = \int Y_2^2 dA \tag{7-30a}$$

and, when Equation (7-29b) is substituted into this expression, it becomes

$$I_{Y11} = \cos^2\theta \int X_2^2 dA + \sin^2\theta \int X_1^2 dA - 2\sin\theta\cos\theta \int X_1 X_2 dA$$

The integrals in this equation are the moments and products of inertia given by Equations (7-26) and (7-27). Therefore, the expression may be written:

$$I_{Y11} = I_{X11}\cos^2\theta + I_{X22}\sin^2\theta + 2I_{X12}\sin\theta\cos\theta \tag{7-30b}$$

The same type of procedure can be followed with the moment of inertia I_{Y22} and the product of inertia I_{Y12} in order to obtain the following results:

$$I_{Y22} = \int Y_1^2 dA \tag{7-31a}$$

or

$$I_{Y22} = I_{X11}\sin^2\theta + I_{X22}\cos^2\theta - 2I_{X12}\sin\theta\cos\theta \tag{7-31b}$$

and

$$I_{Y12} = -\int Y_2 Y_1 dA \tag{7-32a}$$

or

$$I_{Y12} = -I_{X11}\sin\theta\cos\theta + I_{X22}\sin\theta\cos\theta + I_{X12}(\cos^2\theta - \sin^2\theta) \tag{7-32b}$$

Equations (7-30b), (7-31b), and (7-32b) give the moments and products of inertia of the area with respect to the second system of axes in terms of I_{X11}, I_{X22}, and I_{X12} for the first system. Thus, if the latter quantities and the angle θ are known, the inertia quantities for the rotated axes can be calculated without difficulty.

In order to show the rotation properties of moments and products of inertia, it is convenient to handle them as a matrix. Therefore, the *inertia matrix* $\mathbf{J}_X$ for the first system of axes is formulated with the moments of inertia on the main diagonal and the products of inertia in the off-diagonal positions:

$$\mathbf{J}_X = \begin{bmatrix} I_{X11} & I_{X12} \\ I_{X21} & I_{X22} \end{bmatrix} \tag{7-33}$$

Similarly, the inertia matrix for the second system of axes is

$$\mathbf{J}_Y = \begin{bmatrix} I_{Y11} & I_{Y12} \\ I_{Y21} & I_{Y22} \end{bmatrix} \tag{7-34}$$

The matrices $\mathbf{J}_X$ and $\mathbf{J}_Y$ are square and symmetric, and they are related by the following rotation transformation:

$$\mathbf{J}_Y = \mathbf{R}\mathbf{J}_X\mathbf{R}^{\mathrm{T}} \tag{7-35}$$

in which $\mathbf{R}$ is the two-dimensional rotation matrix. Equation (7–35) can be verified by substituting $\mathbf{J}_X$ into the right-hand side and performing the matrix multiplications. It will be found that the product $\mathbf{R}\mathbf{J}_X\mathbf{R}^\mathrm{T}$ produces the matrix $\mathbf{J}_Y$, the elements of which are given by Equations (7–30b), (7–31b), and (7–32b). Thus, the inertia matrix is transformed from the first system of coordinates to the second system by an orthogonal similarity transformation using the rotation matrix $\mathbf{R}$ as the transformation matrix. It should be recalled that the rows of $\mathbf{R}$ contain the direction cosines for the second system of axes with respect to the first system. Equation (7–35) provides an efficient means of calculating moments and products of inertia for the second system of axes when the corresponding quantities are known for the first system of axes. The reverse transformation is

$$\mathbf{J}_X = \mathbf{R}^\mathrm{T}\mathbf{J}_Y\mathbf{R} \tag{7–36}$$

as can be seen by comparison with Equation (7–20).

The inertia matrix can be diagonalized by rotating axes to principal axes, as explained in Section 7.3. The rotation matrix $\mathbf{R}_P$ used for this purpose is equal to the transpose of the orthogonal modal matrix $\mathbf{Q}$ for the matrix to be diagonalized (see Equation 7-21). In this case, the matrix to be diagonalized is $\mathbf{J}_X$, and the application of Equation (7–23) gives

$$\mathbf{J}_P = \mathbf{Q}^\mathrm{T}\mathbf{J}_X\mathbf{Q} = \begin{bmatrix} I_{P11} & 0 \\ 0 & I_{P22} \end{bmatrix} \tag{7–37}$$

in which $\mathbf{J}_P$ is the *principal inertia matrix*. The diagonal elements I_{P11} and I_{P22} are the *principal moments of inertia* for the principal axes p_1 and p_2, respectively (see Figure 7-5b). These inertia quantities are defined as follows:

$$I_{P11} = \int P_2^2 dA \qquad I_{P22} = \int P_1^2 dA \tag{7–38}$$

in which P_1 and P_2 are the coordinates of the element of area dA with respect to the principal axes. From the discussion of Section 7.3, it is apparent that I_{P11} and I_{P22} are the eigenvalues of the inertia matrix $\mathbf{J}_X$. The products of inertia of the area A are zero for the principal axes:

$$I_{P12} = I_{P21} = -\int P_2 P_1 dA = 0 \tag{7–39}$$

An important property of the principal moments of inertia can be obtained from the characteristic equation of the inertia matrix $\mathbf{J}_Y$. This equation is (see Equation 7–34):

$$(I_{Y11} - \lambda)(I_{Y22} - \lambda) - I_{Y12}^2 = 0 \tag{a}$$

or when expanded,

$$\lambda^2 - (I_{Y11} + I_{Y22})\lambda + I_{Y11}I_{Y22} - I_{Y12}^2 = 0 \tag{b}$$

The values of λ that are the roots of these equations are the principal moments of inertia I_{P11} and I_{P22}. The characteristic polynomial on the left-hand sides of Equations (a) and (b) is a function of λ. This characteristic function will be denoted $f(\lambda)$, as follows:

$$f(\lambda) = (I_{Y11} - \lambda)(I_{Y22} - \lambda) - I_{Y12}^2 \tag{c}$$

or

$$f(\lambda) = \lambda^2 - (I_{Y11} + I_{Y22})\lambda + I_{Y11}I_{Y22} - I_{Y12}^2 \tag{d}$$

The values of λ that satisfy the characteristic equation (Equation a or b) are those values that make $f(\lambda)$ equal to zero. Consider now how $f(\lambda)$ varies for different values of λ. If λ is a very large number (either positive or negative), the first term in the polynomial in Equation (d) predominates, and $f(\lambda)$ is seen to be a large positive number. Next, suppose that λ is equal to either I_{Y11} or I_{Y22}; then the first term in the polynomial in Equation (c) becomes zero, and $f(\lambda)$ is equal to $-I_{Y12}^2$. This value is always negative because the square of I_{Y12} must be positive. Since both I_{Y11} and I_{Y22} are positive, as shown by their definitions (Equations 7-30a and 7-31a), it follows that the graph of the quadratic function $f(\lambda)$ has the general shape shown in Figure 7-6. Thus, the two values of λ that satisfy the characteristic equation are such that one is always larger than the two moments of inertia I_{Y11} and I_{Y22}, and the other is always smaller than these two quantities. Since I_{Y11} and I_{Y22} are the moments of inertia for an arbitrary set of axes y_1 and y_2, the important conclusion is reached that the principal moments of inertia [the intercepts for $f(\lambda) = 0$ in Figure 7-6] are the maximum and minimum moments of inertia of the area A for all axes through the same origin.

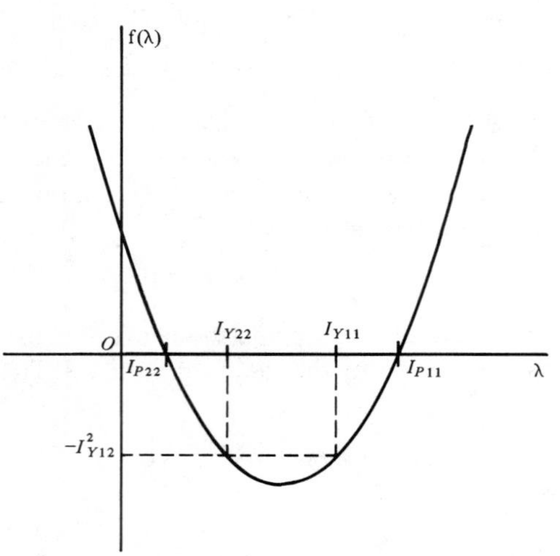

Figure 7-6. Graph of characteristic function

Example. The calculation of moments and products of inertia for rotated axes, including the determination of the principal moments of inertia, will be illustrated using the rectangular area shown in Figure 7-7. The rectangle has sides of lengths a and b, and the origin of coordinates is taken at the lower left-hand corner. Axes x_1 and x_2 coincide with two edges of the rectangle.

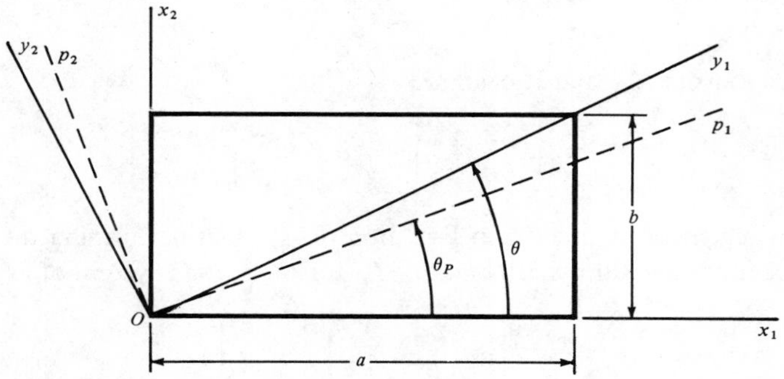

Figure 7-7. Example: Rectangular area

The inertia properties of the area with respect to the x_1x_2 axes are determined from Equations (7-26) and (7-27):

$$I_{X11} = \int X_2^2 dA = \int_0^b \int_0^a X_2^2 dX_1 dX_2 = \frac{ab^3}{3}$$

$$I_{X22} = \int X_1^2 dA = \int_0^b \int_0^a X_1^2 dX_1 dX_2 = \frac{a^3 b}{3}$$

$$I_{X12} = -\int X_2 X_1 dA = -\int_0^b \int_0^a X_2 X_1 dX_1 dX_2 = -\frac{a^2 b^2}{4}$$

From these results the inertia matrix for the x_1x_2 axes is obtained:

$$\mathbf{J}_X = \frac{ab}{12}\begin{bmatrix} 4b^2 & -3ab \\ -3ab & 4a^2 \end{bmatrix}$$

Now assume that the inertia matrix $\mathbf{J}_Y$ is to be determined for a second set of axes that are aligned with the diagonal of the rectangle (y_1y_2 axes in the figure). The angle θ between the x_1 and y_1 axes is such that

$$\cos\theta = \frac{a}{c} \qquad \sin\theta = \frac{b}{c}$$

in which

$$c = \sqrt{a^2 + b^2}$$

Therefore, the rotation matrix is

$$\mathbf{R} = \frac{1}{c}\begin{bmatrix} a & b \\ -b & a \end{bmatrix}$$

In order to have a more specific example, assume now that the length a is equal to twice the length b:

$$a = 2b$$

Then the matrices $\mathbf{J}_X$ and $\mathbf{R}$ become

$$\mathbf{J}_X = \frac{b^4}{3}\begin{bmatrix} 2 & -3 \\ -3 & 8 \end{bmatrix} \qquad \mathbf{R} = \frac{1}{\sqrt{5}}\begin{bmatrix} 2 & 1 \\ -1 & 2 \end{bmatrix}$$

Substituting these matrices into Equation (7-35) and performing the matrix multiplications gives the inertia matrix $\mathbf{J}_Y$ for the second system of axes:

$$\mathbf{J}_Y = \mathbf{R}\mathbf{J}_X\mathbf{R}^\mathrm{T} = \frac{b^4}{15}\begin{bmatrix} 4 & 3 \\ 3 & 46 \end{bmatrix}$$

Thus, the moments and products of inertia for the y_1y_2 axes are

$$I_{Y11} = \frac{4b^4}{15} \qquad I_{Y22} = \frac{46b^4}{15} \qquad I_{Y12} = \frac{b^4}{5}$$

The same steps can be followed if the inertia matrix $\mathbf{J}_Y$ is to be calculated for any other set of axes having a known orientation with respect to the x_1x_2 axes.

As a special case, the axes can be rotated through an angle equal to 90°; the rotation matrix in this instance becomes

$$\mathbf{R} = \begin{bmatrix} 0 & 1 \\ -1 & 0 \end{bmatrix}$$

and Equation (7-35) gives

$$\mathbf{J}_Y = \mathbf{R}\mathbf{J}_X\mathbf{R}^\mathrm{T} = \frac{b^4}{3}\begin{bmatrix} 8 & 3 \\ 3 & 2 \end{bmatrix}$$

This calculation shows that when the axes are rotated 90°, the two moments of inertia are interchanged, and the product of inertia changes its sign.

Consider now the task of determining the principal axes and principal moments of inertia for the rectangle (for which $a = 2b$). The eigenvalues of the matrix $\mathbf{J}_X$ can be found as described in Section 6.1; the results are

$$\lambda_1 = \frac{b^4}{3}(5 + 3\sqrt{2}) = 3.081b^4$$

$$\lambda_2 = \frac{b^4}{3}(5 - 3\sqrt{2}) = 0.252b^4$$

The eigenvectors also can be found as described in Chapter 6, and they are

$$\mathbf{X}_1 = \beta_1 \begin{bmatrix} 1-\sqrt{2} \\ 1 \end{bmatrix} \qquad \mathbf{X}_2 = \beta_2 \begin{bmatrix} 1+\sqrt{2} \\ 1 \end{bmatrix}$$

When these vectors are normalized, they become

$$\mathbf{x}_1 = \begin{bmatrix} -0.383 \\ 0.924 \end{bmatrix} \qquad \mathbf{x}_2 = \begin{bmatrix} 0.924 \\ 0.383 \end{bmatrix}$$

Since the vectors $\mathbf{x}_1$ and $\mathbf{x}_2$ represent the direction cosines of the principal axes, it is apparent that the second of these axes will lie in the first quadrant of the original axes because both direction cosines are positive. At this point, it is advantageous to recall that the sequence in which eigenvalues are arranged is arbitrary. Therefore, the values of λ_1 and λ_2 can be switched (and similarly for the corresponding vectors $\mathbf{x}_1$ and $\mathbf{x}_2$) in order to cause the principal axis corresponding to the first eigenvector to lie in the first quadrant. Thus, restating the results in this manner, the eigenvalues become

$$\lambda_1 = 0.252b^4 \qquad \lambda_2 = 3.081b^4$$

and the rearranged modal matrix is

$$\mathbf{Q} = \begin{bmatrix} 0.924 & -0.383 \\ 0.383 & 0.924 \end{bmatrix}$$

The rotation matrix $\mathbf{R}_P$ which corresponds to the modal matrix is (see Equation 7-21):

$$\mathbf{R}_P = \mathbf{Q}^\mathrm{T} = \begin{bmatrix} 0.924 & 0.383 \\ -0.383 & 0.924 \end{bmatrix}$$

A comparison of this matrix with the 2 × 2 rotation matrix (Equation 7-13) shows that $\cos\theta_P = 0.924$ and $\sin\theta_P = 0.383$. Therefore, $\theta_P = 22.5°$, and this angle is indicated in Figure 7-7. Thus, principal axes p_1 and p_2 are rotated 22.5° from the original set of axes,* and the corresponding principal moments of inertia of the rectangle are $0.252b^4$ and $3.081b^4$, respectively. These conclusions may be confirmed by evaluating Equation (7-37), as follows:

$$\mathbf{J}_P = \begin{bmatrix} 0.924 & 0.383 \\ -0.383 & 0.924 \end{bmatrix} \frac{b^4}{3} \begin{bmatrix} 2 & -3 \\ -3 & 8 \end{bmatrix} \begin{bmatrix} 0.924 & -0.383 \\ 0.383 & 0.924 \end{bmatrix}$$

$$= \begin{bmatrix} 0.252b^4 & 0 \\ 0 & 3.081b^4 \end{bmatrix} = \begin{bmatrix} I_{P11} & 0 \\ 0 & I_{P22} \end{bmatrix}$$

*It should also be recalled from Section 7.3 that there are actually four positions of the axes p_1 and p_2 that can be designated as principal axes. The other three positions are 90°, 180°, and 270° from the position shown in Figure 7-7.

As the $y_1 y_2$ axes in Figure 7-7 are rotated about the origin O, the moments and products of inertia of the area are changing continuously. When the axes assume the position of the principal axes, the products of inertia become zero, and the moments of inertia are the minimum and maximum values shown above.

As a matter of interest, the traces of the matrices $\mathbf{J}_X$, $\mathbf{J}_Y$, and $\mathbf{J}_P$ can be calculated:

$$\text{tr}(\mathbf{J}_X) = I_{X11} + I_{X22} = \frac{2b^4}{3} + \frac{8b^4}{3} = \frac{10b^4}{3}$$

$$\text{tr}(\mathbf{J}_Y) = I_{Y11} + I_{Y22} = \frac{4b^4}{15} + \frac{46b^4}{15} = \frac{10b^4}{3}$$

$$\text{tr}(\mathbf{J}_P) = I_{P11} + I_{P22} = 0.252b^4 + 3.081b^4 = 3.333b^4$$

All of these values are the same, because the trace of a matrix is not changed by a similarity transformation. The same conclusion may be obtained from Equations (7-30b) and (7-31b) by observing that

$$I_{Y11} + I_{Y22} = I_{X11} + I_{X22} \tag{7-40}$$

which means that the sum of the moments of inertia remains constant as the axes are rotated through any angle θ.

The inertia quantities for a volume are defined in a manner analogous to those for a plane area. For instance, the moment of inertia of a volume with respect to any axis is defined as

$$I = \int \rho^2 dV \tag{e}$$

in which dV represents the volume of an infinitesimal element (see Figure 7-8) and ρ is the distance from the element to the axis under consideration. If the moment of inertia of the volume with respect to the axis x_1 is desired, then the square of the distance ρ_1 from the element to the axis is

$$\rho_1^2 = X_2^2 + X_3^2 \tag{f}$$

Similarly, for the x_2 axis the square of the distance is

$$\rho_2^2 = X_1^2 + X_3^2 \tag{g}$$

and for the x_3 axis it is

$$\rho_3^2 = X_1^2 + X_2^2 \tag{h}$$

Substitution of Equations (f), (g), and (h) into Equation (e) produces the three moments of inertia of the volume with respect to the $x_1 x_2 x_3$ axes:

$$I_{X11} = \int (X_2^2 + X_3^2) dV \qquad I_{X22} = \int (X_1^2 + X_3^2) dV$$
$$I_{X33} = \int (X_1^2 + X_2^2) dV \tag{7-41}$$

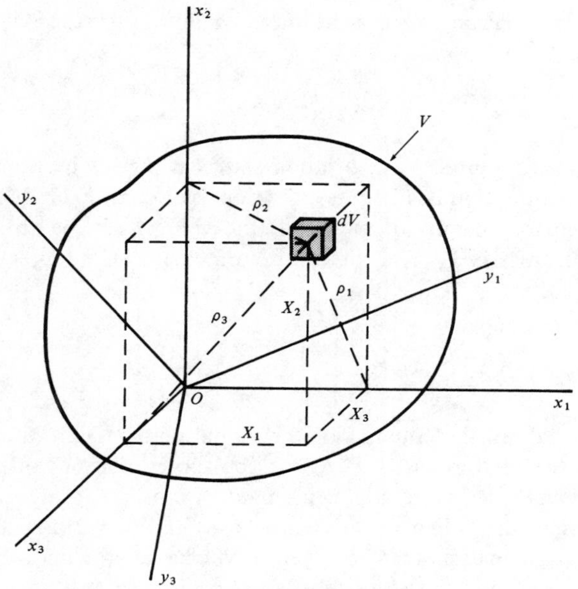

Figure 7-8. Volume

The products of inertia of a volume are defined as follows:

$$\begin{aligned} I_{X12} &= -\int X_2 X_1 dV & I_{X21} &= -\int X_1 X_2 dV \\ I_{X13} &= -\int X_3 X_1 dV & I_{X31} &= -\int X_1 X_3 dV \\ I_{X23} &= -\int X_3 X_2 dV & I_{X32} &= -\int X_2 X_3 dV \end{aligned} \tag{7-42}$$

It is apparent from these definitions that

$$I_{X12} = I_{X21} \qquad I_{X13} = I_{X31} \qquad I_{X23} = I_{X32} \tag{7-43}$$

The inertia quantities in Equations (7–41) and (7–42) may be placed into a 3×3 inertia matrix $\mathbf{J}_X$ with the moments of inertia on the main diagonal and the products of inertia in the off-diagonal positions; thus:

$$\mathbf{J}_X = \begin{bmatrix} I_{X11} & I_{X12} & I_{X13} \\ I_{X21} & I_{X22} & I_{X23} \\ I_{X31} & I_{X32} & I_{X33} \end{bmatrix} \tag{7-44}$$

Similarly, the inertia matrix $\mathbf{J}_Y$ for a second set of axes $y_1 y_2 y_3$ (see Figure 7-8) may be constructed:

$$\mathbf{J}_Y = \begin{bmatrix} I_{Y11} & I_{Y12} & I_{Y13} \\ I_{Y21} & I_{Y22} & I_{Y23} \\ I_{Y31} & I_{Y32} & I_{Y33} \end{bmatrix} \tag{7-45}$$

Then the matrix equation relating the inertia matrices for the two systems of axes is the same as for areas:

$$\mathbf{J}_Y = \mathbf{R}\mathbf{J}_X\mathbf{R}^\mathsf{T} \tag{7-35}$$
repeated

except that now the required 3 × 3 rotation matrix **R** is given by Equation (7-4). The reverse transformation is, of course, the same as Equation (7-36).

The principal moments of inertia of the volume can be obtained as the eigenvalues of the matrix $\mathbf{J}_X$, and rotation to principal axes is represented by the formulation

$$\mathbf{J}_P = \mathbf{Q}^\mathsf{T}\mathbf{J}_X\mathbf{Q} = \begin{bmatrix} I_{P11} & 0 & 0 \\ 0 & I_{P22} & 0 \\ 0 & 0 & I_{P33} \end{bmatrix} \tag{7-46}$$

In this transformation, the columns of the modal matrix **Q** are the normalized eigenvectors of the inertia matrix $\mathbf{J}_X$. It can be shown that one of the principal moments of inertia will be the largest moment of inertia for any system of rotated axes through the origin O (see Figure 7-8) and one will be the smallest. The remaining principal moment of inertia will be intermediate between the other two. However, it is also possible for two, or even all three, of the principal moments of inertia to be equal. In a case where all three moments of inertia are equal, any set of orthogonal axes through the same origin will be principal axes. In other words, the matrices $\mathbf{J}_X$, $\mathbf{J}_Y$, and $\mathbf{J}_P$ will all be diagonal matrices, and the modal matrix **Q** will be equal to the identity matrix.

The concepts of moments and products of inertia and their transformation properties under rotation of axes have great usefulness in physical analyses. In dynamics, the mass moments and products of inertia of a material body play an important role. The expressions for these quantities are similar to the corresponding quantities for volumes, except that the element of volume dV is replaced by an element of mass dM. With this change in the definitions of the inertia terms, the inertia matrices for the mass of a solid will be given by Equations (7-44) and (7-45), and their transformations will be given by Equations (7-35) and (7-46).*

7.5. Rotation of Axes in *n* Dimensions. The concepts of rotation of axes for vectors and matrices and transformation to principal coordinates can be extended by analogy from two and three dimensions to n dimensions. To do so requires the introduction of an orthogonal coordinate system $x_1, x_2, \ldots, x_n$ in

---

*The inertia matrices are analogous mathematically to other quantities of engineering interest, such as the stress and strain matrices encountered in theory of elasticity. Quantities such as the inertia properties of a body that transform under rotation of axes according to Equation (7-35) are called *tensors*. Thus, one can speak of the inertia tensor, the stress tensor, or the strain tensor. A discussion of the general properties of tensors is beyond the scope of this book.

n-dimensional space. Thus, a vector **V** of n dimensions can be represented in the $x_1x_2...x_n$ coordinate system as the vector **X**:

$$\mathbf{X} = \begin{bmatrix} X_1 \\ X_2 \\ \cdots \\ X_n \end{bmatrix} \tag{7-47}$$

The elements of this vector are the components of **V** in the $x_1x_2 \ldots x_n$ system of coordinates. The coordinate system is assumed to be orthogonal, which means that the unit vectors corresponding to all components are mutually orthogonal.

The vector **V** may also be represented in a second orthogonal coordinate system $y_1, y_2, \ldots, y_n$ of n dimensions. The transformation matrix **T** that relates the vector **Y** in the second system to the vector **X** in the first system is an orthogonal matrix of nth order that can be interpreted as a generalized rotation matrix. The matrix equation for such a transformation is

$$\mathbf{Y} = \mathbf{TX} \tag{7-48a}$$

and the reverse transformation is

$$\mathbf{X} = \mathbf{T}^{-1}\mathbf{Y} = \mathbf{T}^{\mathrm{T}}\mathbf{Y} \tag{7-48b}$$

By analogy with the three-dimensional case, the rows of the transformation matrix **T** may be considered to consist of the direction cosines of the axes in the y system with respect to the axes in the x system. Successive rotation of vectors in n-dimensional space can be carried out in the same manner as for three-dimensional vectors (see Section 7.1).

The transformation of matrices between the first and second coordinate systems follows the pattern developed for rotation of axes in two and three dimensions. Thus, if two n-dimensional vectors $\mathbf{X}_1$ and $\mathbf{X}_2$ in the first system are related as follows:

$$\mathbf{AX}_1 = \mathbf{X}_2 \tag{7-49}$$

then the transformation to the second system is accomplished by the equations

$$\mathbf{Y}_1 = \mathbf{TX}_1 \qquad \mathbf{Y}_2 = \mathbf{TX}_2 \tag{7-50a}$$

$$\mathbf{B} = \mathbf{TAT}^{-1} = \mathbf{TAT}^{\mathrm{T}} \tag{7-50b}$$

$$\mathbf{BY}_1 = \mathbf{Y}_2 \tag{7-50c}$$

The reverse transformations are

$$\mathbf{X}_1 = \mathbf{T}^{-1}\mathbf{Y}_1 = \mathbf{T}^{\mathrm{T}}\mathbf{Y}_1 \qquad \mathbf{X}_2 = \mathbf{T}^{-1}\mathbf{Y}_2 = \mathbf{T}^{\mathrm{T}}\mathbf{Y}_2 \tag{7-51a}$$

$$\mathbf{A} = \mathbf{T}^{-1}\mathbf{BT} = \mathbf{T}^{\mathrm{T}}\mathbf{BT} \tag{7-51b}$$

The transformations represented by Equations (7-50b) and (7-51b) are orthogonal similarity transformations; therefore, the matrices **A** and **B** have the same eigenvalues, the same trace, and the same determinant. Furthermore, the eigenvectors of **B** are related to those of **A** by Equation (7-48a).

Transformation to principal coordinates $p_1, p_2, \ldots, p_n$ for a symmetric matrix **A** is carried out in the same manner as above, except that the transpose of the modal matrix **Q** of the normalized eigenvectors of **A** is used in place of the matrix **T**. The transformation is represented by the equations

$$\mathbf{P}_1 = \mathbf{Q}^{\mathrm{T}}\mathbf{X}_1 \qquad \mathbf{P}_2 = \mathbf{Q}^{\mathrm{T}}\mathbf{X}_2 \tag{7-52a}$$

$$\mathbf{Q}^{\mathrm{T}}\mathbf{A}\mathbf{Q} = \mathbf{S} \tag{7-52b}$$

$$\mathbf{S}\mathbf{P}_1 = \mathbf{P}_2 \tag{7-52c}$$

in which $\mathbf{P}_1$ and $\mathbf{P}_2$ are n-dimensional vectors consisting of the components of $\mathbf{X}_1$ and $\mathbf{X}_2$ in the principal coordinate system, and **S** is the spectral matrix. The last equation shows that the ratios of corresponding elements of the vectors $\mathbf{P}_2$ and $\mathbf{P}_1$ are equal to the eigenvalues of **A**. Thus, it can be seen that all of the matrix operations used in rotation of axes in three-dimensional space have their counterparts when dealing with vectors and matrices in n dimensions. These operations are extremely useful in many types of engineering problems, such as vibration analysis of multi-degree-of-freedom elastic systems.

7.6. Nonorthogonal Coordinate Transformations. In the preceding section, the transformation of vectors and matrices from one orthogonal coordinate system to another was summarized. It was assumed throughout that discussion that the transformation matrix **T** was orthogonal. If the matrix **T** is not orthogonal, but instead is any square nonsingular matrix, then the following vector transformations

$$\mathbf{Y} = \mathbf{T}\mathbf{X} \tag{7-53a}$$

and

$$\mathbf{X} = \mathbf{T}^{-1}\mathbf{Y} \tag{7-53b}$$

can no longer be considered to be rotation of axes from one orthogonal coordinate system to another. Instead, they are said to represent *linear transformations*, of which rotation of axes is a special case. If desired, a linear transformation can be considered as a change of coordinate systems, except that the rows of **T** are not unit vectors in the directions of the new axes.

In addition to vector transformations, nonorthogonal similarity transformations of matrices can be performed also. For this purpose, Equations (7-49) through (7-51) may be generalized to include nonorthogonal transformations by omitting from those equations the expressions containing the transposed matrix $\mathbf{T}^{\mathrm{T}}$.

If the matrix **A** has n linearly independent eigenvectors, then it is similar to a diagonal matrix (see Section 6.5), and a modal matrix **M** can be used to transform **A** to diagonal form. Thus, the equations for transforming to principal coordinates (see Equations 7-52) become

$$\mathbf{P}_1 = \mathbf{M}^{-1}\mathbf{X}_1 \qquad \mathbf{P}_2 = \mathbf{M}^{-1}\mathbf{X}_2 \tag{7-54a}$$

$$\mathbf{M}^{-1}\mathbf{A}\mathbf{M} = \mathbf{S} \tag{7-54b}$$

$$\mathbf{S}\mathbf{P}_1 = \mathbf{P}_2 \tag{7-54c}$$

in which $\mathbf{P}_1$ and $\mathbf{P}_2$ are n-dimensional vectors.

The preceding equations provide a means of solving simultaneous equations by transforming to principal coordinates and diagonalizing the coefficient matrix. In order to demonstrate this method, assume that the simultaneous equations are expressed in the matrix form:

$$\mathbf{A}\mathbf{X} = \mathbf{C} \tag{7-55}$$

in which **A** is a nonsingular matrix of coefficients having n linearly independent eigenvectors, **X** is a vector of n unknowns, and **C** is a vector of right-hand sides. Assume also that a modal matrix **M** and the corresponding spectral matrix **S** for the matrix **A** have been determined. The vectors **X** and **C** transform to principal coordinates according to Equations (7-54a):

$$\mathbf{P}_1 = \mathbf{M}^{-1}\mathbf{X} \qquad \mathbf{P}_2 = \mathbf{M}^{-1}\mathbf{C} \tag{7-56}$$

and the matrix **A** transforms by Equation (7-54b). Therefore, in the principal coordinates the simultaneous equations take the form of Equation (7-54c). Solving for $\mathbf{P}_1$ in that equation gives

$$\mathbf{P}_1 = \mathbf{S}^{-1}\mathbf{P}_2 \tag{7-57}$$

Substituting $\mathbf{P}_1$ and $\mathbf{P}_2$ from Equations (7-56) and Equation (7-57) yields

$$\mathbf{M}^{-1}\mathbf{X} = \mathbf{S}^{-1}\mathbf{M}^{-1}\mathbf{C}$$

and premultiplying both sides by **M** results in

$$\mathbf{X} = \mathbf{M}\mathbf{S}^{-1}\mathbf{M}^{-1}\mathbf{C} \tag{7-58}$$

Equation (7-58) represents a direct solution for the vector of unknowns. A comparison of this equation with Equation (7-55) shows that the premultipliers of **C** must be equal to the inverse of **A**, which is the same result that was derived previously in Equation (6-33):

$$\mathbf{A}^{-1} = \mathbf{M}\mathbf{S}^{-1}\mathbf{M}^{-1} \qquad \text{(6-33) repeated}$$

Thus, the transformation to principal coordinates and the resulting diagonalization of the coefficient matrix is equivalent to matrix inversion. The above method of solving equations is useful if the eigenvalues and eigenvectors of the coefficient matrix have been determined beforehand for some other purpose. This situation is encountered in vibrations problems, where the natural frequencies (eigenvalues) and mode shapes (eigenvectors) of a system are usually determined initially and then are available for use in subsequent analyses. The use of Equation (7-58) is simplified somewhat if the matrix **A** is symmetric, because then the orthogonal matrix **Q** can be taken as the modal matrix, and its inverse is replaced by its transpose.

Selected References

1. Hohn, F.E., *Elementary Matrix Algebra*, 3rd ed., Macmillan, New York, 1973, 522 pages. (This mathematics textbook is an excellent general reference for engineers on matrices, determinants, inversion, simultaneous equations, eigenvalue problems, and more advanced topics of linear algebra.)

2. Hadley, G., *Linear Algebra*, Addison-Wesley, Reading, Massachusetts, 1961, 290 pages. (Another excellent book on the same subjects as Hohn, above.)

3. Bronson, R., *Matrix Methods–An Introduction*, Academic Press, New York, 1969, 284 pages. (Still another excellent book on the same subjects as Hohn, above.)

4. Fuller, L.E., *Basic Matrix Theory*, Prentice-Hall, Englewood Cliffs, New Jersey, 1962, 245 pages. (A good textbook on matrices for the beginner.)

5. Frazer, R.A., Duncan, W.J., and Collar, A.R., *Elementary Matrices and Some Applications to Dynamics and Differential Equations*, Cambridge University Press, New York, 1955, 416 pages. (A widely referenced classical textbook on the title subject.)

6. Faddeev, D.K., and Faddeeva, V.N., *Computational Methods of Linear Algebra*, W.H. Freeman, San Francisco, 1963, 621 pages. (This comprehensive book treats in detail the various computational schemes for evaluating determinants, inverting matrices, solving simultaneous equations, and finding eigenvalues and eigenvectors; many numerical examples are solved.)

7. Jennings, A., *Matrix Computation for Engineers and Scientists,* Wiley, London, 1977, 330 pages. (This excellent book describes computational techniques for matrix multiplication, solution of simultaneous linear equations, solution of eigenvalue problems, etc. It is intended for use by those who plan to write their own computer programs.)

8. Scarborough, J.B., *Numerical Mathematical Analysis,* 6th ed., The Johns Hopkins University Press, Baltimore, 1966, 600 pages. (An excellent text on numerical methods of computation; includes a chapter on solving simultaneous linear equations by determinants, elimination, inversion of matrices, and iteration; many solved examples are given.)

Answers to Problems

Chapter 1

1.3-1 $\mathbf{A} + \mathbf{B} = [-3.8\ \ 7.5\ \ 3.0\ \ -0.2]$
$\mathbf{C} - \mathbf{D} = \{4.1\ \ 1.7\ \ 5.8\ \ 1.3\}$

1.3-2 $\mathbf{R}_1 = [43.6\ \ -36.5\ \ -3.0\ \ 8.9]$
$\mathbf{R}_2 = \{3.7\ \ 3.6\ \ -18.6\ \ 27.5\}$

1.3-3 $\mathbf{R}_3 = [-11\ \ 8\ \ 19\ \ -4\ \ 4]$

1.3-4 $\mathbf{R}_4 = \{0.4\ \ -1.9\ \ 24.4\ \ -26.2\}$

1.4-1 $\mathbf{A} \cdot \mathbf{B} = -131.73$
$\mathbf{C} \cdot \mathbf{D} = 14.92$

1.4-2 $\mathbf{R} = [1095\ \ 0\ \ 1241\ \ 292\ \ -1460]$

1.4-3 $A = 8.58\quad C = 4.84$

1.4-4 -4.36

1.4-5 $\alpha = 135.1°$

1.4-6 -7.88

1.4-8 $\mathbf{A} \times \mathbf{B} = -\mathbf{B} \times \mathbf{A} = [-11\ \ 14\ \ 54]$

1.5-1 $x = -6.0$

1.5-2 $x_1 = 6\quad x_2 = -10$

1.5-3 $x_1 = 2\quad x_2 = -6$

1.5-4 $x_1 = 0\quad x_2 = 4\quad x_3 = 24$

1.5-5 $\mathbf{a} = [0.605\ \ -0.086\ \ 0.346\ \ 0.173\ \ -0.691]$
$\mathbf{b} = [0.099\ \ 0.198\ \ 0.891\ \ 0\ \ -0.396]$

1.5-6 $\mathbf{e}_1 = [0.272\ \ 0.953\ \ 0.136]$

1.5-7 $\mathbf{e}_2 = [0.213\ \ 0.426\ \ -0.853\ \ -0.213]$

Chapter 2

2.2-1 $$\mathbf{C} = \begin{bmatrix} -2 & 3 \\ -2 & 5 \\ -8 & -7 \end{bmatrix}$$

2.2-2 $$\mathbf{C} = \begin{bmatrix} 16 & -3 \\ 11 & 15 \end{bmatrix}$$

2.4-1 $\mathbf{AB} = \{14\ \ 29\}$

2.4-2 $\mathbf{AB} = \{-4\ \ 29\ \ -8\}$

2.4-3 $\mathbf{BC} = \begin{bmatrix} 6 & -6 & 9 \\ -4 & 4 & -6 \\ 2 & -2 & 3 \\ 8 & -8 & 12 \end{bmatrix}$

2.4-4 $\mathbf{DE} = \begin{bmatrix} 4 & -5 \\ 45 & 37 \end{bmatrix}$ $\mathbf{ED} = \begin{bmatrix} 15 & 3 & 0 & 36 \\ 1 & -2 & -2 & -6 \\ -28 & 12 & 16 & 0 \\ 5 & 1 & 0 & 12 \end{bmatrix}$

2.4-6 $\mathbf{AE} = \begin{bmatrix} 26 & 21 \\ 26 & -4 \end{bmatrix}$

2.4-7 $\mathbf{EA} = \begin{bmatrix} 33 & -6 & 0 & 15 \\ -4 & 0 & 2 & -1 \\ -12 & 8 & -16 & -12 \\ 11 & -2 & 0 & 5 \end{bmatrix}$

2.4-8 $\mathbf{FG} = \begin{bmatrix} 3 & 7 \\ -7 & -8 \end{bmatrix}$ $\mathbf{GF} = \begin{bmatrix} -10 & 15 \\ -5 & 5 \end{bmatrix}$

2.4-9 $\mathbf{HJ} = \mathbf{0}$ $\mathbf{JH} = \begin{bmatrix} 0 & 0 & 0 \\ 0 & 0 & 0 \\ 33 & 15 & 0 \end{bmatrix}$

2.5-3 $\mathbf{D}_1\mathbf{D}_2 = \mathbf{D}_2\mathbf{D}_1 = \begin{bmatrix} -8 & 0 & 0 \\ 0 & 9 & 0 \\ 0 & 0 & -1 \end{bmatrix}$

2.5-7 $\mathbf{D}_1^3 = \begin{bmatrix} 8 & 0 & 0 \\ 0 & 27 & 0 \\ 0 & 0 & -1 \end{bmatrix}$

2.5-8 $\mathbf{A}_1^3 = \begin{bmatrix} -45 & 12 & 12 \\ -24 & 9 & -24 \\ -42 & 12 & -27 \end{bmatrix}$

2.7-1 $\mathbf{AB} = \begin{bmatrix} 2 & -3 \\ 0 & 3 \end{bmatrix}$

2.7-2 $\mathbf{CD} = \begin{bmatrix} 10 & 7 & 3 & 1 \\ -10 & 3 & 2 & 1 \\ 10 & -1 & -11 & 4 \\ -5 & 9 & -4 & 1 \end{bmatrix}$

2.7-3 $\mathbf{EF} = \begin{bmatrix} 8 & 4 & -6 & 7 \\ 10 & -4 & 1 & 1 \\ 0 & -2 & -6 & 5 \end{bmatrix}$

2.7-5 $\mathbf{G}^2 = \begin{bmatrix} 1 & 2 & 2 & 0 & 3 \\ 6 & 5 & 2 & 4 & 6 \\ 0 & 2 & 1 & 0 & 2 \\ 4 & 4 & 1 & 3 & 4 \\ 4 & 4 & 0 & 4 & 3 \end{bmatrix}$

2.8-1 $\mathbf{AB} = \begin{bmatrix} 2 & 11 \\ 1 & 4 \end{bmatrix}$

2.8-2 $\mathbf{CD} = \begin{bmatrix} 25 & 20 \\ 30 & 40 \end{bmatrix}$

2.8-5 $\mathbf{A}^3 = \begin{bmatrix} 20 & 21 \\ 7 & 6 \end{bmatrix}$

Chapter 3

3.2-2 $D_1 = 25$

3.2-3 $D_2 = 39$

3.2-4 $D_3 = 26xyz$

3.2-5 $x = 17$

3.2-6 $b = 2a = -2/3$

3.2-7 $D_6 = 110$

3.2-8 $D_7 = 968$

3.2-9 $|\mathbf{U}| = A_{11}A_{22}A_{33}\ldots A_{nn}$

3.3-5 $|\mathbf{EFG}| = 96$

3.4-1 $D_1 = -19$

3.4-5 $D_2 = 4326$

3.4-6 $D_3 = -951$

Chapter 4

Note: Because the matrix inverses can be checked by multiplying with the original matrices, only a few answers are given.

4.4-2

$$\mathbf{A}^{-1} = \begin{bmatrix} 4 & -4 & 1 \\ 14 & -13 & 3 \\ -21 & 21 & -5 \end{bmatrix}$$

4.4-3 First row of $\mathbf{B}^{-1}$: [46 151 17]

4.4-4

$$\mathbf{C}^{-1} = \frac{1}{22}\begin{bmatrix} 16 & -9 & -2 \\ -20 & 3 & 8 \\ -10 & 7 & 4 \end{bmatrix}$$

4.4-6 First row of $\mathbf{E}^{-1}$: $\frac{1}{35}[29 \quad -15 \quad 17]$

4.4-8

$$\mathbf{G}^{-1} = \begin{bmatrix} 1 & 0 & 0 \\ -\lambda_1 & 1 & 0 \\ \lambda_1\lambda_3 - \lambda_2 & -\lambda_3 & 1 \end{bmatrix}$$

4.4-9 Last row of $\mathbf{H}^{-1}$: $\frac{1}{6}[-3 \quad 4 \quad -2 \quad 2]$

4.4-10 First row of $\mathbf{J}^{-1}$: $\frac{1}{6}[5 \quad -4 \quad 3 \quad -2 \quad 1]$

4.4-12 $\mathbf{L}^{-1} = \mathbf{L}$

4.6-4 $A_{12} = -A_{21} = \pm\frac{\sqrt{3}}{2}$

4.6-5 Yes

4.6-6 Yes

4.6-7 $x_1 = -0.769$
$x_2 = -0.547$

4.7-6

$$\mathbf{F}^{-1} = \frac{1}{14}\begin{bmatrix} 6 & 6 & 16 \\ 2 & 9 & 17 \\ 2 & 2 & 10 \end{bmatrix}$$

4.7-7 First row of $\mathbf{G}^{-1}$: [0.534 −0.397 0.080]

4.7-8
$$\mathbf{H}^{-1} = \begin{bmatrix} 0.0785 & -0.0317 & -0.0146 \\ -0.0204 & 0.0540 & -0.0163 \\ -0.0280 & -0.0033 & 0.0621 \end{bmatrix}$$

4.7-9 First row of $\mathbf{J}^{-1}$: [0.00306 0.00224 −0.00300]

4.7-11 First row of $\mathbf{K}^{-1}$: [0.945 0.645 0.013]

4.7-12 $\mathbf{L}^{-1} = \mathbf{L}$

4.7-13 First row of $\mathbf{M}^{-1}$: [0.0617 0.0463 −0.0127 −0.0208]

4.7-14 Last row of $\mathbf{N}^{-1}$: [5.46 −2.45 −0.75 1.24]

4.7-15 First row of $\mathbf{P}^{-1}$: $\frac{1}{42}$ [13 −10 7 −4 1]

4.7-16 First row of $\mathbf{Q}^{-1}$: [0.289 0.068 −0.332 −0.162]

4.7-17 First row of $\mathbf{R}^{-1}$: [0 −1 −1 0 1 1]

4.8-5
$$\mathbf{C}^{-1} = \begin{bmatrix} 1.7 & 2.9 & 0.8 \\ 1.5 & 2.6 & 1.6 \\ 0.6 & 1.2 & 2.3 \end{bmatrix}$$

4.8-7 First row of $\mathbf{D}^{-1}$: [0.8 −0.6 0.4 −0.2]

4.8-9 First row of $\mathbf{E}^{-1}$: [0.5607 0.0405 −0.0026 0.0370]

Chapter 5

Note: Because the solutions of the simultaneous equations can be checked by substitution into the original equations, only a few answers are given.

5.2-2 $\mathbf{X} = \{-3 \quad 4\}$

5.2-4 $\mathbf{X} = \{2 \quad -3 \quad -2\}$

5.2-6 $\mathbf{X} = \{-6 \quad 2 \quad 5\}$

5.4-4 $\mathbf{X} = \{3 \quad -7 \quad 10\}$

5.4-6 $\mathbf{X} = \{14.20 \quad -29.70 \quad 31.50\}$

5.4-8 $\mathbf{X} = \{5 \quad -2 \quad 3 \quad 7\}$

5.4-10 $\mathbf{X} = \{2 \quad -1 \quad 5 \quad 5\}$

5.4-12 $\mathbf{X} = \{-2 \quad 3 \quad 1 \quad -4 \quad 5\}$

5.5-2 $\mathbf{X} = \beta\{17 \quad 6 \quad 43\}$

5.5-4 $\mathbf{X} = \beta\{-5 \quad 2 \quad 2\}$

5.5-6 $\mathbf{X} = \beta\{5 \quad -2 \quad -1 \quad 1\}$

5.6-1 Dependent

5.6-2 Dependent

5.6-3 Independent

5.6-4 Dependent

5.6-5 $x = 10$

5.6-6 $\mathbf{F}_1 = 2\mathbf{F}_2 - \mathbf{F}_3 + \mathbf{F}_4$

5.6-7 $\mathbf{G}_1 = -\frac{1}{3}\mathbf{A}_2 + \frac{4}{3}\mathbf{A}_3$

5.7-1 One

5.7-2 $r(\mathbf{I}_n) = n$

5.7-3 One; one

5.7-4 $r(\mathbf{A}) = 3 \quad r(\mathbf{B}) = 2 \quad r(\mathbf{C}) = 1$

5.7-5 $r(\mathbf{D}) = 3 \quad r(\mathbf{E}) = 2$

5.8-3 $\mathbf{X} = \beta\{2 \quad 5\} + \{0.6 \quad 0\}$

5.8-4 $\mathbf{X} = \mathbf{O}$

5.8-5 $\mathbf{X} = \beta\{1 \quad -6\}$

5.8-6 (a) $\frac{a}{d} = \frac{b}{e}$ and $\frac{a}{d} \neq \frac{c}{f}$

(b) $\frac{a}{d} \neq \frac{b}{e}$

(c) $\frac{a}{d} = \frac{b}{e} = \frac{c}{f}$

5.8-7 $\mathbf{X} = \{1 \quad -1 \quad 2\}$

5.8-8 $\mathbf{X} = \beta\{2 \quad 13 \quad 5\} + \{1.4 \quad 2.6 \quad 0\}$

5.8-9 $\mathbf{X} = \beta_1\{2 \quad 1 \quad 0\} + \beta_2\{4 \quad 0 \quad 1\} + \{3 \quad 0 \quad 0\}$

5.8-10 No solution

5.8-11 No solution

5.8-12 $\mathbf{X} = \mathbf{O}$

5.8-13 $\mathbf{X} = \beta\{-5 \quad 7 \quad 13\}$

5.8-14 $\mathbf{X} = \beta_1\{3 \quad -2 \quad 0\} + \beta_2\{1 \quad 0 \quad 2\}$

Chapter 6

Note: Because the solutions for the eigenvalues and eigenvectors can be checked by substitution into Equation (6-1), only a few answers are given.

6.1-1 $(1 - \lambda)^3$

6.1-3 $\lambda_1 = 65 \quad \lambda_2 = 26$

6.1-4 $\lambda_1 = 10 \quad \lambda_2 = 4 \quad \lambda_3 = -2$

6.2-2 $\lambda_1 = 7 \quad \mathbf{X}_1 = \beta_1\{1 \quad 1\}$
$\lambda_2 = 2 \quad \mathbf{X}_2 = \beta_2\{1 \quad 2\}$

6.2-4 $\lambda_1 = 13 \quad \mathbf{X}_1 = \beta_1\{1 \quad 0\}$
$\lambda_2 = 2 \quad \mathbf{X}_2 = \beta_2\{0 \quad 1\}$

6.2-6 $\lambda_1 = 4 \quad \mathbf{X}_1 = \beta_1\{-1 \quad 1 \quad 1\}$
$\lambda_2 = 2 \quad \mathbf{X}_2 = \beta_2\{1 \quad 0 \quad 2\}$
$\lambda_3 = 1 \quad \mathbf{X}_3 = \beta_3\{0 \quad 0 \quad 1\}$

6.2-8 $\lambda_1 = 5 \quad \mathbf{X}_1 = \beta_1\{2 \quad 5 \quad -3\}$
$\lambda_2 = 3 \quad \mathbf{X}_2 = \beta_2\{3 \quad 4 \quad 0\}$
$\lambda_3 = -1 \quad \mathbf{X}_3 = \beta_3\{1 \quad 2 \quad -1\}$

6.3-2 $\lambda_1 = 2 \quad \mathbf{X}_1 = \beta_1\{4 \quad 5 \quad 0\}$
$\lambda_2 = 2 \quad \mathbf{X}_2 = \beta_2\{1 \quad 0 \quad -5\}$
$\lambda_3 = -3 \quad \mathbf{X}_3 = \beta_3\{3 \quad 6 \quad 8\}$

6.3-4 $\lambda_1 = 1 \quad \mathbf{X}_1 = \beta_1\{1 \quad 0 \quad 0\}$
$\lambda_2 = 1 \quad \mathbf{X}_2 = \beta_2\{1 \quad 0 \quad 0\}$
$\lambda_3 = 1 \quad \mathbf{X}_3 = \beta_3\{1 \quad 0 \quad 0\}$

6.4-2 $\lambda_1 = 3 \quad \mathbf{X}_1 = \beta_1\{1 \quad 1 \quad 0\} \quad \mathbf{X}_1 = \beta_1\{1 \quad 0 \quad -2\}$
$\lambda_2 = 1 \quad \mathbf{X}_2 = \beta_2\{0 \quad 1 \quad 0\} \quad \mathbf{X}_2 = \beta_2\{-1 \quad 1 \quad 2\}$
$\lambda_3 = -1 \quad \mathbf{X}_3 = \beta_3\{2 \quad 0 \quad 1\} \quad \mathbf{X}_3 = \beta_3\{0 \quad 0 \quad 1\}$

6.4-4 $\lambda_1 = 5$ and $1/5 \quad \mathbf{X}_1 = \beta_1\{1 \quad 0\}$
$\lambda_2 = -4$ and $-1/4 \quad \mathbf{X}_2 = \beta_2\{2 \quad 1\}$

6.4-6 $\lambda_1 = 9 \quad \mathbf{X}_1 = \beta_1 \{1 \quad -2 \quad 0\}$
$\lambda_2 = 6 \quad \mathbf{X}_2 = \beta_2 \{1 \quad -1 \quad -1\}$
$\lambda_3 = 3 \quad \mathbf{X}_3 = \beta_3 \{0 \quad 0 \quad 1\}$

6.4-8 $\lambda_1 = 1 \quad \mathbf{X}_1 = \beta_1 \{\sqrt{3} \quad 1\}$
$\lambda_2 = -1 \quad \mathbf{X}_2 = \beta_2 \{1 \quad -\sqrt{3}\}$

6.4-10 $\lambda_1 = 1 \quad \mathbf{X}_1 = \beta_1 \{1 \quad 1 \quad 0\}$
$\lambda_2 = 1 \quad \mathbf{X}_2 = \beta_2 \{\sqrt{2} \quad 0 \quad 1\}$
$\lambda_3 = -1 \quad \mathbf{X}_3 = \beta_3 \{-1 \quad 1 \quad \sqrt{2}\}$

6.4-12 $\lambda_1 = 6 \quad \mathbf{X}_1 = \beta_1 \{32 \quad 16 \quad 5\}$
$\lambda_2 = 0 \quad \mathbf{X}_2 = \beta_2 \{0 \quad 10 \quad 1\}$
$\lambda_3 = -10 \quad \mathbf{X}_3 = \beta_3 \{0 \quad 0 \quad 1\}$

6.4-14 $\lambda_1 = d_1 \quad \mathbf{X}_1 = \beta_1 \{1 \quad 0 \quad 0\}$
$\lambda_2 = d_2 \quad \mathbf{X}_2 = \beta_2 \{0 \quad 1 \quad 0\}$
$\lambda_3 = d_3 \quad \mathbf{X}_3 = \beta_3 \{0 \quad 0 \quad 1\}$

6.7-2 $\lambda_1 = 26 \quad \mathbf{X}_1 = \beta_1 \{4 \quad 0 \quad 7\}$
$\lambda_2 = 17 \quad \mathbf{X}_2 = \beta_2 \{0 \quad 1 \quad 0\}$
$\lambda_3 = -39 \quad \mathbf{X}_3 = \beta_3 \{7 \quad 0 \quad -4\}$

6.7-4 $\lambda_1 = 50 \quad \mathbf{X}_1 = \beta_1 \{0 \quad 0 \quad 1\}$
$\lambda_2 = 0 \quad \mathbf{X}_2 = \beta_2 \{3 \quad 4 \quad 0\}$
$\lambda_3 = -75 \quad \mathbf{X}_3 = \beta_3 \{4 \quad -3 \quad 0\}$

6.7-6 $\lambda_1 = 8 \quad \mathbf{X}_1 = \beta_1 \{1 \quad 1 \quad 0\}$
$\lambda_2 = 8 \quad \mathbf{X}_2 = \beta_2 \{0 \quad 0 \quad 1\}$
$\lambda_3 = -2 \quad \mathbf{X}_3 = \beta_3 \{-1 \quad 1 \quad 0\}$

6.7-8 $\lambda_1 = 20 \quad \mathbf{X}_1 = \beta_1 \{1 \quad -1 \quad 0\}$
$\lambda_2 = 0 \quad \mathbf{X}_2 = \beta_2 \{1 \quad 1 \quad 0\}$
$\lambda_3 = 0 \quad \mathbf{X}_3 = \beta_3 \{0 \quad 0 \quad 1\}$

6.7-10 $\lambda_1 = 12 \quad \mathbf{X}_1 = \beta_1 \{1 \quad 1 \quad \sqrt{2}\}$
$\lambda_2 = 4 \quad \mathbf{X}_2 = \beta_2 \{1 \quad -1 \quad 0\}$
$\lambda_3 = 4 \quad \mathbf{X}_3 = \beta_3 \{\sqrt{2} \quad 0 \quad -1\}$

6.8-4
$$\sqrt{\mathbf{A}} = \frac{1}{5}\begin{bmatrix} 4+\sqrt{6} & -2+2\sqrt{6} & 0 \\ -2+2\sqrt{6} & 1+4\sqrt{6} & 0 \\ 0 & 0 & 5\sqrt{3} \end{bmatrix}$$

6.8-5 See Problem 2.5-10.

6.8-6
$$\sqrt{\mathbf{A}} = \begin{bmatrix} 1.4142 & 0 & 0 \\ 0 & 2.8981 & 1.2654 \\ 0 & 1.2654 & 1.8436 \end{bmatrix}$$

6.8-7
$$\mathbf{A}^5 = \begin{bmatrix} 197 & -66 \\ 495 & -166 \end{bmatrix}$$

6.8-8
$$\mathbf{A}^3 = \begin{bmatrix} 44 & 86 & 0 \\ 86 & 173 & 0 \\ 0 & 0 & 27 \end{bmatrix}$$

6.9-1 $\lambda_1 = 9.979 \quad \mathbf{X}_1 = \beta_1 \{1 \quad 2.220 \quad 1.539\}$
$\lambda_2 = 3.389 \quad \mathbf{X}_2 = \beta_2 \{1 \quad 0.070 \quad -0.751\}$
$\lambda_3 = 2.632 \quad \mathbf{X}_3 = \beta_3 \{1 \quad -1.290 \quad 1.212\}$

6.9-2 $\lambda_1 = 20.59 \quad \mathbf{X}_1 = \beta_1 \{1 \quad 0.7697 \quad 0.6414\}$
$\lambda_2 = -6.118 \quad \mathbf{X}_2 = \beta_2 \{1 \quad -10.32 \quad 10.83\}$
$\lambda_3 = 3.524 \quad \mathbf{X}_3 = \beta_3 \{1 \quad -0.6811 \quad -0.7417\}$

6.9-3 $\lambda_1 = 1150.8 \quad \mathbf{X}_1 = \beta_1 \{1 \quad 1.351 \quad 1.033\}$

6.9-4 $\lambda_1 = 58.41 \quad \mathbf{X}_1 = \beta_1 \{1 \quad 1.113 \quad 0.695\}$

Index